The Careless Driver

(AKA The Distracted Driver)

By

Chaz Van Heyden

With updated statistics and chapter on

The DUI Driver.

Velvet Gloves Publishing
Nashville, Tenn

© 2017 Velvet Gloves Library
2nd Edition © 2018-2019

ALL RIGHTS RESERVED

PRINTED IN THE UNITED STATES

DEDICATED TO THE DRIVERS
WHO DID GET TRAINED
ADEQUATELY

Reviews

"Van Heyden does a marvelous job with *The Careless Driver* of making us think in regards to the state of our driving in today's society. With an obvious disregard to others on the roadway, we have forgotten the common courtesy of the system. It's amazing to me to see, through interview, how our education has altered over the years. This is a book that is needed in the hands of all potential licensees." John Blanks (Technical Sales for Tesla)

"An interesting read for people learning to drive. It's a book that is a good reference text that will be most beneficial to be ear marked and reread many times. Loaded with solid information for drivers of all ages. This book should be made available to students of all driving schools. If it prevents even one fatality on the roadways, then Mr. Van Heyden made a huge accomplishment here. Add this one to your library, give one to a friend learning to drive, make highways safer. Congrats on another fine endeavor" F. Botham (Author)

"I really enjoyed *The Careless Driver*. The author does a good job of getting multiple perspectives on driving education or lack there of and sheds light on some of the flaws in the American driver education system. The book has large text and is easy to read."

C. Moore (Banking Officer)

I was very curious about the book and its contents, and I was very pleased to see that someone else noticed some of the strange driving behaviors out on the road. It was news to me how little education there really is for new drivers. You can kind of tell if you spend any time on the road some of these behaviors could be prevented with enough knowledge and teaching, and some of the normal behaviors of driving on the road. I agree with the conclusions in the book, especially hands-on experience is always the best. If [state] legislators get copies of this book there is a chance they will introduce bills along the lines recommended in *The Careless Driver*. It would be a good direction to go in. Brandon Williams-Business Owner

Contents

Preface	9
Interviews with Driver Training Schools	15
National Highway Traffic Safety Admin. Data	27
Initial Interviews with Drivers	31
Interview with Lee with data from Japanese friends	43
Interviews with Texas and Central India driver	51
Interviews-Amy (AL) through Tim (NC) to Levi (OH)	61
Interview with Rajat from India and Ryan (Tenn)	69
Interview-John (KS) / Thomas (TX) Elaine, Chris/Rajiv	73
Interview with Trevor (Buffalo, NY)	85
Interview with Geeta and Glen	91
Interview with Carlos from the Philippines	103
Interview-Allison (North N.J.) & Marisa (Chicago)	109
Interview with Patrick from Rwanda	119
The DUI Driver & The Texting Driver (2nd Edition)	131
Author's summation & conclusions plus additional statistics	137

PREFACE

We as a country have sunk into a complacency about vehicle death and homicide on our roadways, while many states require little in the way of driving skills to obtain a driver's license.

Can anything be done about the yearly 35,000 plus deaths of humans caused by inattentive and unskilled drivers?

Few, if any of the vehicle drivers that irritate, annoy and scare you do it with malice. What I've observed is that they <u>drive inexpertly</u>.

It is unlikely that anyone has taken the time and effort to observe, analyze *directly and record* the driving habits and practices *from* drivers to the extent this work does. This may be the first. Hopefully the data about them will assist Driver Training Schools, The State Departments of Public Health (DMVs), parents of teenagers, traffic courts, and give drivers across America who read this book an opportunity to see where we have erred in our insufficient *driver training*, and where we can lower the stress substantially with and in the rush hour Interstate or highway traffic conditions due to the presence of under-educated, under-trained drivers. Let's now define what a Careless Driver is.

DEFINITION Careless Driver: Distracted driver or an impaired driver, this leads to a remedy more suitable and more quickly. This includes a driver who is not one hundred percent there when driving a vehicle. Note: This last sentence, in body, is at the very end of lengthy descriptions in state codes about the 'negligent' and 'careless' driver with some states omitting it entirely.

For myself, I have chauffeured thousands of individuals. And, while doing so have had the opportunity to drive in all types of weather and roads and traffic conditions, seeing all types of driver *driving* weaknesses and errors—Airport shuttle in Los Angeles and the ride-shares in Nashville and surrounding cities for more than four years. Having great familiarity with driving in cities, rural towns and on Interstates and seeing the enormous amount of poor driving, daily, I decided to launch a grass-roots survey into driver *driving habits and practices,* as well as their training or lack of it, and see if there was anything to learn. I decided to interview at random a number of my passengers/riders, several from other countries and states, and compile their candid stories of how and when they first learned to drive, and any further driver training that they may have experienced. Let's see whether or not there <u>is something that can be done to get driving back to a pleasure again.</u>

    The following interviews were done throughout the Spring and Summer of 2017 with passengers in my chauffeured vehicle. They represent a factual cross-section of types of drivers, not only from Tennessee but other parts of the world. The following 26 plus interviews with drivers and 2 interviews with major city owner / instructors of Driver Training Schools should prove eye-opening and in many cases very surprising. Of the 28 interviews, several are of naturalized citizens living here in the U.S., from India, Kazakhstan, the Philippines, Nepal and Japan. We will see as we read through that these countries see driver training in a much different light, and why. Highway deaths had been steadily dropping to approximately 35,000 compared with earlier totals of 50,000 plus. But 2016 has seen a substantial increase, now the deadliest year in nearly a decade. (Read online article: fortune.com/2017/02/15/traffic-deadliest-year) Statistics for roadway fatalities 2017 will not be available until December 2018! The truly sad part of these statistics is *too* many gifted artists have been taken from us by automobile crashes: Patti Santos (age 40) of *It's A Beautiful Day*, Jimmie Spheeris (Motorcycle-age 34), Duane Allman (age 24), Steve Allen, noted comic and TV host. But why settle for even 35,000 deaths or permit thousands of deaths and a million times that amount in damaged bodies, hospitalized bodies and ruined property? Why? That is what this work also seeks to spotlight, but with a positive answer. It should be known that the Graduated Driver License (GDL) program of many states has been the only effective program, up to 30% reduction in fatalities, but again why stop there with only *beginner drivers?* **What if we as a nation expensed a small fraction per capita (per driver) of the billions spent yearly for auto insurance, but on professional driver training?**

    The author's questions, responses and comments <u>are underlined for easy recognition,</u> but the answers, though verbatim from the recordings, are not in quotations for simpler assimilation and conveyance. (Meaning transcribing, editing / proofreading time and effort.) The recorded interviews are in relatively sequential date order. Statements / words in parentheses ( ) are to the reader. First names only are used. Some, only a few, had difficulty in recalling data that was asked for in the interview, and this is sometimes footnoted or noted directly in parentheses also. I did try to not repeat the same questions, mainly since in most cases there would not be enough time for the passenger to answer a battery of questions. Also, to get at a variety of scenarios and driving practices polled. And, sometimes there will be a "End." at the end of an interview, when it is not completely apparent that the interview is finished.

As a footnote: the driver errors on surface streets observed by myself when driving more than thirty thousand miles a year, combined with the errors and misinstructions found in the interviews that follow are intensified where they are observed on Interstates, where the flow of cars and trucks/buses is substantially faster requiring more skill and *alertness* than on city or rural roadways.

Usually, there are more miles of non-Interstate roadways than Interstate roadways in any given state. And, the Interstates are mostly used by motorists during the day hours and at *rush hour*. So, it is telling that for Tennessee and most states (2016 data), there are more deaths per road mile on Interstates than on other major and minor *arteries,* as they are referred to by the National Center for Statistics & Analysis (National Highway Traffic Safety Administration – NHTSA)

The remedies proposed in later chapters by no means constitute the sole resolution to significantly lowering the incidence of deaths and injuries on our roadways occasioned by unalert, distracted, impaired and under-trained drivers. As a Nashville Councilman who has read this book stated: "[Our city] is working on reducing traffic fatalities by lowering speed limits and evaluating our most dangerous intersections. This looks like an additional part of the solution."

The Careless Driver
(The Under-trained Driver)

Pitner Driving School (Memphis, TN) provides six hours of practical driving instruction of the thirty hours in the course that it delivers. Their main client is the parents of youths about to get a car, ages 16, 17, 18. They are, by the owner Mr. Pitner's perception, of Pitner Driving School, that his is the top of several driving/traffic schools in Memphis, Tennessee. Here I want to again make a distinction between traffic schools (that instruct about the rules of the road and safe driving practices) and *driving schools* that give the hands-on training to new adult and young drivers so that they are better equipped to handle the fast-paced Interstate driving situations and dangers.

Note: This interview, was not conducted with a recording as were all the driver / rider interviews done first hand and recorded. Notes were taken as the questions posed were answered.

Further, Mr. Pitner went on to say that his instructors take their students out onto the Interstate and educate them in how to merge correctly with the faster moving traffic, and this done repetitively until the student can do it confidently. "We won't take the student out when it is raining or snowing," stated Mr. Pitner but we plan for doing that practical instruction on another schedule if the weather doesn't permit us to deliver the hands-on training due to inclement weather conditions. Although the driving school's curriculum is not posted on the Internet, each day the student goes out with the instructor, notes are taken and brought back to be typed up by the school's secretary and presented to the parents of the student after the thirty-hour training.

<u>Phil</u>:

who is an instructor for Brentwood Driver Training Program (Brentwood – near Nashville, TN): "[We're] the only completely dual control, we have steering wheels, we have brake pedals, we have gas pedals. Other schools only have brake pedals. [Those that] take our class get four hours behind the wheel instruction, or six hours more, two of which are behind simulators. We're open to persons purchasing lessons outside of that. So, I've had people get as many as twelve two-hour lessons. Or as few as one. It's mainly adults. The ones that buy that many is because they move here maybe from out of country and don't have the support structure to practice as much.

With the teenage students, typically their parents drive with them more.

We take them (students) out anywhere from 9:00 AM to 7:00 PM, Monday through Saturday. I've had students who have driven in rain storms.

<u>Do you repeatedly take (students) out to the Interstate to teach them to merge with traffic properly?</u> I do as much as I can. I tell them to wait until the dotted line portion (ends and not [cross] the solid line. And, tell them to not rush through it. Use their signal, look for an opening before you go. Try to get your speed up, before you go. Try to get your speed up to the speed of traffic before you get over. I try to do it multiple times. Sometimes I'm limited by traffic. If I'm in rush hour traffic and it's a two-hour lesson I may be able to merge on maybe once, twice at the most. Most of us [our instructors] try to get our students on the Interstate at least once during a lesson. You may not get the full opportunity to merge like more than once but most of us try to take them on the Interstate at least one time.

*Note: Phil states, "Try" in answering but it is not the optimum level of skill needed. Drivers of Interstates <u>must</u> merge smoothly or slow-downs and collisions result.

<u>What about following too closely?</u> We teach them the two-second rule. Because it increases the distance based on speed. Something we always teach in our driving school, following safely.

Changing lanes from the slower to the faster lanes as you go left. We teach them not to break while shifting lanes—from left lanes to right hand lanes. That sometimes you must speed up. But, you want to base what you're doing on the traffic. If the cars next to you are going over your speed, you know faster, you want to accelerate into that lane.

<u>How do you teach them to know that, to be aware of that?</u> You can gauge how fast drivers are going and have their pedal usage reflect what they're seeing in their mirrors. <u>How successful do you think you are with your students?</u> I'd like to be more successful but most of the time if they have two lessons with me, *then I'll see a noticeable difference in their driving. (That's the standard?)*

<u>Have you seen anybody stopping on the Interstate merge lane with no Yield sign?</u> I haven't had students do

that because I have pedals on my car, but I've noticed-obviously I won't let them do that. I've seen other drivers do that while with my students. It's a great learning opportunity to point it out later.

<u>Tell me about looking ahead.</u> We teach them to keep their eyes elevated where the two lines on the side of the road come together in the distance to a focal point, and their hand eye coordination helps them to stay between the lines. I tell my students it's a similar (inaudible word) like when you throw a baseball to someone you don't look at the ground in front of you. You look at them and your body just gets it there.

<u>What about when they get behind the Escalades and the Avalanches (wide SUVs)?</u> Tell them to create more distance, so they can keep their eyes off without getting so close to the car in front of them. You must teach them how to pay attention to the cars in front of them without focusing too much on just that car. So, see the brake lights, see what's happening but while keeping your eyes elevated.

<u>Here's the million-dollar question-do you teach any of your students that the gears are not just for accelerating but they are also for decelerating?</u> Not really. It doesn't

come up that often. I mainly just teach them to brake whenever they are going down a hill, something like that.

I mainly teach my students to use the brake but that's primarily so that the drivers behind us will know we are braking and slowing down, because with a student driver car they are much more likely to hit their brake too hard or just *brake at the wrong time.* (???) I try to get the drivers behind us alerted so that we are slowing down.

Have you taken any students out in a parking lot to show them how to handle a skid? It comes up in the lesson all the time. I don't typically take them into a parking lot to show them that, but I'll take them into a neighborhood and drive around and show them what happens if they accelerate too hard or brake too hard. It comes up at least one time, every time it rains.

What about turns on the Interstate—do they really know how to take the turn correctly, because I see a lot of drivers braking into the turns instead of braking before the turn. Yeah, I teach them where to brake, stuff like that. And where to accelerate.

One more thing—emergency vehicles, actually I've asked many people, for inclusion in this book, and I have noticed invariably, when there is a collision, even if it on the other side of the Interstate [drivers] slowing down. Yeah (Phil chuckles),

It's a rubberneck*. What are you trying to do, cause another accident and slow traffic down even more? Yeah, I tell my students if it's not on your side of the Interstate there really is no reason to overreact to it. Read about it in the news later.

*rub·ber·neck [ˈrəbərˌnek]
_Verb: turn one's head to stare at something in a foolish manner: a passerby rubbernecking at the accident scene.

"a passerby rubbernecking at the accident scene" Oxford Dictionaries © Oxford University Press

Here is an example of the effort being made in local areas to improve driving habits: DEFENSIVE DRIVING CLASS: On Saturday December 9, from 9 a.m. to 3 p.m. there will be a New York State approved Defensive Driving Class. Save 10% on your base auto insurance for the next three years & receive up to 4 points off your driving record according to New York State Department of Motor Vehicle guidelines. The class will be held at First Baptist Church, 45 Washington Street, Saratoga Springs. Fee $35 Bring a friend $30 each. A portion of the fee will be donated to First Baptist Church. Registration is required and can be made by calling Ray Frankoski at 518-286-3788.

As we can see the incentives are skewed towards saving $ on insurance and vacating or nullifying points against one's driver's record. Ideally, persons attending should be attending to acquire skills resulting in improvement as *drivers*. Only hands-on *practical instruction* behind the wheel of a vehicle is going to provide that kind of improvement.

These are a few of the many driving faults seen on Nashville Interstate and surface streets and elsewhere:

#1 Following too closely the vehicle immediately in front.

#2 Braking instead of gearing down to slow the vehicle (in conjunction mainly with #1.)

#3 A. Stopping on an Interstate on-ramp merge lane with no Yield sign,

3B. Also merging slowly onto an Interstate highway more slowly than the traffic traveling in the immediate left lane.

#4 Changing lanes abruptly in heavy traffic after only one or two flashes of directional signal.

#5 Slowing down for an Emergency Vehicle with emergency lights flashing when an open lane, two lanes away is available safely.

#6 Slowing down to view an accident on the Interstate* instead of smoothly and with enough notice, changing lanes and getting away from the injury area. (Requires training in looking five to 10 cars ahead) *Including collisions and Emergency vehicles *on the other side of the Interstate middle barrier.*

#7 Left hand turners that turn too early into the oncoming traffic not allowing for the cars ahead that turned to clear the intersection, but blindly following another vehicle ahead and into the path of oncoming vehicles. In other words, misjudging the approaching vehicles in the opposite direction.

#8 A vehicle comes to STOP ON RED, the driver stops and remains there thinking he must stop for red light *and remain there*, even though he is turning right. Also, likewise, on the corner of Bowling Avenue and West End, we have a NO TURN ON <u>FLASHING</u> RED LIGHT, but many times it'll be a red light <u>not flashing</u> and the driver or drivers will be stacked behind each other waiting for who knows what.

#9 Changing lanes on Interstate from slower (right) to faster (left) at speed slower than traffic in left lane.

#10 Looking at only car in front on Interstate (during rush hour especially) instead of looking five to 10 cars ahead, that would alert driver to slow -downs in traffic coming up (most often at Interstate interchanges).

#11 Nashville drivers: unnecessary waiting at red light with four lanes of traffic (2 each) in each direction and *there is no traffic approaching in the nearest lane.*

#12 Again, Nashville drivers, obvious, <u>blatant</u> common error is merging on to fast moving Interstate they will try to maintain speed and not accelerate to merge or even will slow down with several vehicles behind them trying to get on to the Interstate, causing a blockage and potentially a rear-end collision.

#13. (Added after witnessing multiple examples) I've seen drivers close to a vehicle change lanes before signaling or immediately after signaling on the Interstate or surface streets. I've gotten this as a complaint from many passengers not just the passengers I interviewed.

#14. Unawareness by a majority of drivers that lower gears give driver better control of vehicle and a more connected driving experience?

Note: All modern SUV's have sequential gear shifting capability and should be used with the training of how and when to shift *up*. (To a higher rpm gear—which slows the vehicle mechanically, internally instead of burning/wearing out the car's brakes.)

GENERAL DATA ABOUT DEATH RATES-WORLD-WIDE

<From zholpolice.kz [then to /pdd> Excerpt from Russian police. (Be aware this site is sometimes down for routine maintenance and / or problems.)

10.1. The driver must drive the vehicle at a speed not exceeding the specified limit, taking into account traffic volume, characteristics and condition of the vehicle and its cargo, road and weather conditions, in particular the visibility of the direction of travel. <u>Speed should provide the driver increased control over movement of the vehicle to fulfill the requirements of the Rules.</u> (My underline)

The U.S. has seen a 31% reduction in its motor vehicle death rate per capita over the past 13 years. But compared with 19 other wealthy countries, which have declined an average of 56% during the same period, the U.S. has the slowest decrease. Road death rates in countries such as Spain and Denmark have dropped 75.1% and 63.5%, respectively.

If the United States had reduced its death rate to the average of other countries, 18,000 more lives would have been saved, according to the CDC report. cnn.com/2016/07/07/health/us-highest-crash-death-rate.

The title of this book is **The Careless Driver**. However, it is sub-titled *The Under-trained Driver*. And, in this we discover a resolution much easier to accomplish than any other put forward to date. As a wise man observed: "The new model eggbeater or washing machine, the latest year's car, all demand some study and learning before they can be competently operated. When people omit it, there are accidents in the kitchen and piles of bleeding wreckage on the highways." Just *how much learning* is needed with regards to motor vehicles should become by end of this work, strikingly apparent.

INITIAL INTERVIEWS

Joe on 18th May. He learned from his grandfather at age 14, Dixon TN. His mother also instructed him, but she was always nervous in the vehicle. Had no formal training, school or driving school. Talked about speed limits, driving in the rain, driving on the Interstate. Grandfather never took him on the Interstate, only the back roads around the farm. Neither did his father or mother. Father who was paying the insurance, wanted Joe to drive at 5 mph under the speed limit, whereas grandfather said it is OK to drive 5 mph over speed limit. Grandfather taught him at night avoid the glare of oncoming headlights, how to slow down ahead of a curve. Other than that, Joe had very little training about driving on highways and especially super-highways.

Approximate age is 38, he stated he got his license age 16 and that was 22 years ago. First interview for the book *The Careless Driver*.

CJ (African American guy) and I are headed to Walgreens on 5th Ave, Nashville: <u>Who taught you how to drive first?</u> CJ: It was probably split between my parents. Both my mom and my dad. I live out in the country, so if we...in an empty parking lot, or driving through the country, they let me drive where it was uncrowded, unpopulated, so it would be safe. What did they teach you basically in the beginning? Honestly in the beginning, cause it's an adjustment when you first get behind the wheel, on how to actually steer the vehicle and stay on the right side of the rode. How touchy are the brakes and gas. Be prepared to look at your side-view mirrors. How many hours of education from dad and mom you had? Probably over 10 plus hours. Then you got a permit and drove yourself? Then I had to take a class in high school, your sophomore year you take Driver's Ed. too. What did that consist of? That consisted of school, like academic learning from a book, and you got to drive around in a car with your teacher too. Certain days that you went out. Also, you got and a driver's permit allowing that you could drive with anyone over the age of 23. Logging so many hours before you could get your permit. I got my permit in Illinois valley, Cook County. I want to make sure that for the book, when I write it up I'm going to have the people *who are educated here, the people who are educated in another state, and other countries.* I'm going to break it down into three different areas. Good. Because I have interviewed people from Kazakhstan and India. O.K. CJ, when you went out with your instructor you had dual controls, right, what was the main thing he tried to stress with you? Honestly the main thing I remember were the days we went on the highway (Interstate 80). I was really bad when I first was learning to drive. I was trying to be extra cautious. So, when I was merging onto to the highway I would be coasting there, and my speed wouldn't be the right speed to fall in with the other cars. So, he was always stressing that I need to accelerate when I was coming off the ramp and decelerate when coming onto regular roads. That was the biggest thing I took away from that." Was that the Interstate or major highway? Interstate, yeah. I-80, I know it well. Good, Now, did he take you out in the rain? Did he take us out in the rain? Um...I don't know if it ever rained while I was...Well, then they would... That would just be then situational. I know that when I was with my parents, I had driven in the rain before. Oh, really? They were instructing you? Yeah, obviously they were giving me pointers and saying like, don't adjust your turns or anything for the rain, cause if you are going too fast you'll slide. Just little things—it's more common sense when you think about it now, but then it was just make sure you stay calm (garbled few words). Now, on dry pavement did they tell you how to take turns at 60 mph speed limit type highway—you want to go up on a nice little curve...gradually like that (author indicates curve with hands)? It was more just about *now you won't have to check the [steering] wheel over to the side or anything.* I think my dad would always tell me, my dad would say: "When you are driving look at a point ahead of you." "Keep the nose of your car on that point where you are looking. Always have your eyes ahead of you so that way wherever your eyes go your hands will turn with it too." "When you are going on a gradual turn and you have your eyes a little bit in front of you, you'll gradually turn with it instead of looking all the way to the right and just turning your wheel back to the right." Right, good advice. Now, what about rush hour traffic— did your father or instructor ever take you out in *rush-hour* traffic in Illinois? I was from a small town, so I never had rush-hour traffic. So, no, I did not get to experience that, until I went to college out near to Chicago. So, you went to Chicago, big city, lots of traffic; then how did you learn, who taught you there? From there-- It was more or less about me learning by himself. I kinda took things that my parents taught me, while driving around my town. Just applied it to a bigger scale, honestly. The biggest thing with rush hour traffic is staying calm. If—it's like when I was going up there my parents would give me tips before I went. But, I was by myself when I first went into traffic and everything. They said, "If you are overly passive you will get into a wreck too." "You have to hold your own at some point." (Laughing) You know what I mean, don't be overly aggressive, but also don't be overly passive either. Okay, Okay: Did any of your instructors, mother, father, teachers in high school, take you out and try and get you into situation where the car would skid— so you could learn how to handle a skid? Um...yes. My dad, my instructor in school did not. My dad, a driver for UPS. He drives big six axle rigs like that. Right, the long trailer. On Sunday he would pretty much have to drive 35 minutes, so I was driving in the winter too. There were times when there was snow. So, I was in situations where the car could have slid but it didn't. He told me what to expect when the car does start to slide. And, since that time you never have had an actual skid? Oh, yeah, no I've had skids before. How did you handle it? Well, I just went back to what my dad had told me. When your car fishtails turn the wheel the opposite way and it will straighten it back out. (Not the correct method.) If the back end comes out, you want to turn your wheel to the way the tail is going. (That is the correct method.) So, this is to the left, so you turn the wheel to the left. To the left. Yeah. (Laughing) First time I kept spinning in a circle. (Dropped off CJ at convenience store.)

All right it's the 27th of May we are riding with Chandler, to his restaurant work. He's telling me his mom let him drive around in a parking lot-what age were you? (Stated that he first learned about cars and driving from his mom who with him in her lap let him steer in a parking lot. Age 10. Guess mom wanted you to get going quick, huh? That's good. Then, what did she have you do? I wouldn't be in the driver's seat, I was trying to control the steering. (Inaudible) My sister would let me drive around the neighborhood before I even had my permit. So that's how I got started. And, then did you get any formal instruction in High School? No, I didn't take any driver's ed. (Started driving on his own at age 15 down one street to grocery store.) Did you have an uncle or male figure help you out? No, never had anyone. Okay then, when did you start driving on your own?

When I got my permit at 15, I would run errands for my mom like—where we lived at the time we had a grocery store right down the road, so she trusted me not to be able to drive without my permit (Meaning, her in the car.) just down one street. Then once I got my license I just started driving wherever I wanted to go. <u>Okay, okay, very good-and, how would you rate yourself on a scale of 1-10 as a driver here in Nashville?</u> I would say about an "8". Only because I do speed. I'm really bad at speeding sometimes, especially on the Interstate normal rate. <u>Have you been in a car that skidded on wet pavement?</u> No. <u>How far ahead, when you are on the Interstate, are your eyes focused?</u> Pretty far ahead. Of course, I like to see the car in front of me. But, sometimes I can see what I am dealing with if there is no one way ahead of me. Also, it depends on what time of day. <u>How are you at merging into rush-hour traffic?</u> I'm great. I know a lot of people (drivers) don't like it, but I you know I'm one of those drivers that if I know that I have to make a left turn or a right turn I *stay* in the left or right lane. I'm not one of those [drivers] who waits to get over to the right at the last minute. <u>Tell me about Yield Signs.</u> I slow down...I honestly stop at Yield Signs, just because you never know, but in neighborhoods if it's a yield sign I definitely do slow down. O.K. how about stop signs where you can see clearly left and right for two hundred yards? Of course, I stop at a stop sign, do you mean a 4-Way? Let me asks that question again: You're coming to a stop, but you can see there is no traffic visible left or right and no traffic coming your direction? Oh, I see, I still stop. Do you teach anybody else how to drive? Yeah, my step-sister, and she has her own car now, and that's how she got started. And how is she doing? She hasn't wrecked yet...so. I live on my own so Haven't seen her lately. Okay, let me pause this right here. (To locate his drop off place.) So, you think speeding is one of the situations with drivers here, and they are rude. Interesting. And, when you say speeding, what does that consist of? That's a relative term unless we have something to compare it with. Speed limit here is 75 mph, but if I was driving, I would probably be going 60. You have the internal flow of traffic and it is easier (inaudible). Are you aware of any laws or covenants amongst drivers that it is a good idea to drive at the same speed as the flow of traffic? I'm not too sure, but I [drive] not too much faster because that would mean (inaudible). <u>Have you taken any long trips yourself in your vehicle?</u> No. <u>Okay, let's finish it here.</u>

(Caroline visiting Nashville from Atlanta, GA off recorder: says her father taught her the basics of driving a car. Then she had an 8-hour driver's ed. class in high school. Took her out, told her about turn signals, 3-point turns. No, or very little training on Interstate driving.)

(Elizabeth, delivered her to Vanderbilt U. no formal training driving a car. Her friend let Elizabeth use her car in a parking lot. Got her permit in Nashville. No training or instruction from father or mother. No professional instruction.) <u>Just a couple of penetrating questions about your driving habits. How would you rate yourself on a scale of 1-10, on merging onto Interstate rush-hour traffic?</u> Well, I haven't been in an accident... (unintelligible). <u>Describe how you would do it getting onto Interstate traffic Monday morning (rush hour)?</u> I don't know. I don't know what you are asking me. <u>Getting onto the Interstate-do you do that at all?</u> Uh, uh. <u>Okay. Tell me what you are looking for and how you operate the vehicle that gets you onto the Interstate.</u> I don't know what you mean, like? I don't know. <u>Pretend that you are</u>

driving this car onto the Interstate right now. Uh, uh. In rush-hour traffic. How would you tell me to drive? (Considerable hesitation before eventual answer.) Come on! (Forced laugh.) I don't know what you mean, like.... Let's say you're driving this car and not me, okay? And, I'm going to get on the Interstate. Well, I would get into the right lane... Go down the ramp, I don't know. ____ cars. Hm, hmm, what about the speed of the vehicle? Progressively faster as I'm going down the ramp. Do you wait for cars on your left side to go past you? While I'm on the Interstate? While I'm on the ramp? (Said quizzically) On the ramp. Do you wait for cars on your left to go past you? Yes. O.K. Do you slow down to let them go ahead of you? Not if I'm going on the Interstate. Hm, hmm. Have you ever found a situation where a tractor-trailer was on your left and you slowed down or came to a stop? Never a stop...but have to slow down. What was thought process that made you slow down and not speed up ahead of the tractor-trailer? (Typically, tractor-trailers stay out of the right-hand lane, and travel slower than most of the traffic.) It was way bigger than my car. (Laughing) Got it. Can you recall a time when that happened recently? (No.) But it did happen? Sure. O.K. Let's turn this off. End.

Interview with Lee with data
From Japanese friends

It's Friday the 21ˢᵗ of April, and we have Lee (says his name to recorder). And he's native to Japan. Oh, I'm native to Nashville, but I read about stuff from all over the world. (Fact checked later his data as correct) Some of this stuff is from what my friends from Japan have told me. Okay, this is third-hand knowledge, but we'll take it anyway because it's better than any data I've read yet— I'll probably fact-check this. (According to Lee, Japan when certifying or registering drivers, each applicant is expected to spend 2-3000 dollars U.S. on education classes, drive school. Which I find absolutely astonishing, compared to the United States. In addition, new drivers are tagged with a symbol on their windshield[1]1 with a flower, and the flower is called? It's called a Wakaba* mark, a yellow and green sort of thing. It looks like a butterfly; one side or wing is green, and one is yellow. Very distinctive and old people/drivers get a marker and how old do they have to be? I'm going to look at that now...it says 75 and over. All right, that's good—seventy-five and over, O.K. But if they are 70 and they have a

---

1-This sticker or emblem must stay on the driver's windshield for a minimum of one year. More can be learned at https://99percentinvisible.org/article/wakaba-mark-japanese-car-stickers-signal-levels-driving-experience/

condition that could affect their driving, they have to display it too.

How did you come to know this knowledge—are you just interested in looking up things about Japan? That and just asking some Japanese friends. Oh, I see, just in conversation? Right. And, I was just as shocked as you were when I heard when they told me it costs two thousand dollars, at least. Okay we'll look into that. (See fact-check, which I did, and it tallies exactly). Just the fact that you mentioned that this is a practice, this isn't just a one-shot deal, for one person who has had an accident. This is a *practice for all drivers in Japan.* That is astonishing. Yes.

*Wakaba translates to: "Green Leaf"

Yes, because we have our DUI syndrome where you wind up paying $5000.00 to $10,000 and you have to go, if not to prison, you have to go to some kind of community service, and you're probationary for the *rest of your life* (emphasized vocally but not true—only that the DUI never is erased from one's driving record, which was the intent of that remark. Right. So, we have our little scheme for *bad drivers.* (Note: Unfortunately, this is an after-the-act-scheme which is a less than optimum method, since a number of DUI drivers are the cause of roadway fatalities. (Some 10,000 plus, in 2015[2], up from 9, 943 in 2014--of which a percentage are repeat DUI / Alcohol Impaired drivers/offenders.)

<u>We are at two and half minutes of this recording. I think we'll just cut it off. It's all we need...oh, no, I just spotted...what was the name of that flower again?</u> W-a-k-a-b-a. (Lee iterates slowly and carefully.) <u>Let's hear it for Wasabi</u>! (Attempted humor)<u> Lee and author both laughing</u>. (Turned recorder back on) <u>Lee just showed me the symbol for the elderly mark </u>(for an elderly person, 75 years or above). <u>Which is similarly colored but a four-leaf-clover. </u>(Figure #2) <u>Also, very distinctive. So, that wraps it up.</u>

(Compare what your DMV or Public Safety Department requires to license fully a resident in your state compared with the requirements of Japan:)

(Excerpt from online blog regarding applying for Japanese driver license:) "After being issued my new paper license with a 6-month limit, the member of staff at the center explained to me the next step. I needed to go

---

2-Courtesy of our NHTSA at crashstats.nhsta.dot.gov/Api/Public/ViewPublication/812318 -page #19

out on the road at least 5 times with someone who had held a full driver's license for over 3 years. Of course, the car needed to be manual too." – Road Training Section (Read full article at:)

[CITATION Gak12 \l 1033] gakuran.com/driving-in-japan-passing-the-japanese-drivers-test

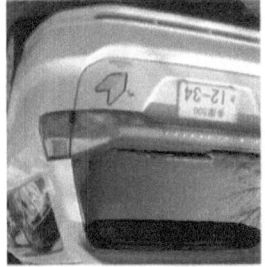  

(Figure 1) (A picture containing clip-art showing a Wakaba: a Japanese emblem meaning "Green Leaf" assigned for one year to new drivers. Green on one half, Yellow on the opposite half. It looks similar to a butterfly and goes on back of car.)

(Figure 2) (A picture containing clip-art showing a Koreisha: a Japanese emblem meaning "Elderly Person" assigned to drivers permanently who are 75 years old and above. In shape of a clover leaf, with four different colors, as a sticker on their automobile.)

(It is interesting and helpful to note that drivers from America and Brazil are required by law in Japan to take a test, to obtain a valid Japanese license, whereas drivers from the United Kingdom and France are not required to be tested.)

gakuran.com/driving-in-japan-passing-the-japanese-drivers-test

(Confidence in American drivers is not high in Japan, where their per capita death rate is less than half of (5.2

per 100,000 persons) compared with America at 11.4 per 100,000.)

http://www.progressive-economy.org/trade_facts/traffic-accidents-kill-1-24-million-people-a-year-worldwide-wars-and-murders-0-44-million

Interviews Texas & E. India

(Jim from Texas states that he drives an SUV. His training in TX did not include instruction on driving in rain or snow. Doesn't use gears in town to slow down vehicle. He mistakenly thought or was instructed that the time to shift to a lower gear is when his vehicle starts down a hill or slope. As for distance maintained behind vehicles on Interstates, he goes by experience, lengthens the usual distance when towing a trailer, and learned the 1 car length per 10 miles per hour rule from instructor.)

We're with Jim, heading toward Bridgestone Arena, downtown Nashville. It's 24th of May 2017, and he's about to answer the one question I ask all my individual riders, which is: Where did you first learn to drive? I think it was Indiana, high school drivers ed. course. Okay. How many practical hours in the seat, behind a moving vehicle did that education afford you? I have no idea. Take a rough guess, a wild guess. 40 hours. O.K. 40 hours, a lot more than usual. Now, do you recall what type of set-up you were in, in the car? Instructor in the passenger seat and three students in back seat. Really? Yeah, and then we rotated. You might find my story a little interesting: I lived in Indiana. Indiana did not have lottery tickets yet. So, every time I drove, I'm assuming since I was a decent driver, we drove to Illinois, so my Instructor could get lottery tickets. How interesting. (Jim is laughing.) And, when I drove he always had a newspaper. One of our four guys drove *everybody paid attention.* Yeah, let's just concentrate on your education...So, were there dual controls in the car? Did he (instructor) have control? He had a brake. Okay, all right. Do you recall any of the instructions he gave you at the beginning that stand out? I don't remember getting much. He just let you drive for practice? Yeah. And, he was just monitoring that? Yup. Did you go on the Interstate? Yeah. Good—did you go on the Interstate during rush hour? Yep. How about during rain? Probably. Okay, probably. That's a 50/50. Yeah. All right, did he teach you how to back the car up correctly? We did, we did parallel parking. (not the same) Parallel parking, no back-up. Did he tell you how to handle a skid? Yeah! He did that. Okay, did he try to simulate that (maneuver) anywhere, even in a parking lot? No. No. O.K. I'm finding most of the instructors never do that. It was in the summer. If it had been in the winter.... Possibly, that would be more appropriate. How about yield and stop signs, did he give you explicit instructions on what they are and how to handle them. Yeah. He did? O.K.

What about following distance, on the Interstate—your car in back of another car? Most of the time he was reading the paper a lot. So, no real instruction on that. Interesting. O.K. Well, this will all go into the book. All right, how would you rate yourself (as a driver) right now on a scale from 1-10, here in Nashville? 9? Good, and why is that? I had a job where I was commuting 150 miles a day. In Chicago. In Chicago. Miles of experience driving in Chicago traffic. Was this a normal car vehicle or was this a truck... No, was a car. Was this in the city proper-with traffic and lights and... Traffic, lights in suburbs...Great. In Indiana, even beyond driver's ed., even before you can get your license they have to have a hundred hours on their own. They have a sheet. Every time they drive we put our initials as a parent. They drove two hours here, three hours there. We must accumulate a hundred hours before they can get their license. Good, good. And, have you done this with your children? And, have you included rush-hour Interstate driving? I'd say some Interstate. Never have then drive in Chicago rush-hour. Okay. My wife would have a heart attack. Now, you are speaking about Chicago as if you— is that your home base? Southwest Indiana. And actually, that Interstate is one of the busiest Interstates in the country. I don't doubt it. So, living here is temporary? I'm on business. End of interview.

So, this is Varsha from India (with heavy accent): which section of India, province? M----------, (E. Indian pronunciation for Central India. Central, that's what I wanted. So, my first question to you is: What are the requirements for a citizen of Central India to get a driver's license? (Her answer begins with an unintelligible set of words-then:) passport, I remember only that. What about your driving skills? Yeah, I have passed road test. Are you required to take any formal training at a training...Yeah, at a training center I have passed my training. Can you briefly describe what training you received at this training school—what type...did they take you out on fast roads like the super-highways. Do they have super-highways in India, or are they just highways? Highways. And, how often do they take you out and train you? Daily, for one month. Do they teach you how to handle a car if it is on a rainy day and it skids? Ya. They did!? Ya. Did they try to make you step on the brake and make the car skid or they told you how to handle it? (Doesn't quite get the question.) Did you skid the car *yourself?* Ya. What was the result? I passed to my test. Okay, let's go back before the skid. When I say skid, I'm talking about a plane, a car starts to move by itself in a certain direction without control. That's a skid. Okay. That's from because the water is on the asphalt. Water is on the road. Okay. And, the car starts to go out of control. That's a skid. You didn't do that? I didn't do that. No. And, in training they didn't tell you what to do when the car begins to go out of control because it's on water? No. No, O.K. Did they tell you how to take a sharp turn? Ya. Okay, did you practice that? Ya. You did. How many times? One time, two times, three times (etc.) I don't remember. O.K., do you think it is less than five times? Ya. Okay, thank you. (Language difference, and her accent definitely a barrier.) (3 minutes 30 seconds into interview) (Pause recording)

What were some of the other skills the instructor taught you about driving? I would have (undecipherable word). (Note: Upon going over the audio of this short section about fifty times, it may be that she replied:) My other skills? (That seems to be her response.) Yeah, other abilities, other duties of a driver—what was the instructor telling you to do as a driver, during this 30-day period?

So...(Varsha tries to recollect...) So, every day you went out, in a car, and an instructor, Ya. Teacher was here (Interviewer points to passenger seat.) Ya. Did they have dual controls? Ya. (She's certain.) Good, so he was taking you through certain neighborhoods and teaching you certain things—what did he teach you mainly? How to drive...Yes, but specifically? What things, how to come to a stop at a stop light, a stop sign? Yeah about a stop line. Okay, did he teach you about speed laws, or how to adjust your speed when it is raining or snowing? When I was taking the training, it was no rain. Okay, thirty days, no rain. All right. What about other traffic, cross-traffic. Did he teach you how to look for oncoming cross traffic at an intersection? Yeah...to see back and forth, right and left, and then I *list speed*. (Here she means I check speed.) Now, when it comes to turning the vehicle on a sharp turn, what did he tell you to do there? Less speed. At what point) did he tell you to do that? Starting of turn. Starting of turn?! (Proper is before turn, and speeding up gradually in the turn so that passengers and driver are not thrown by centrifugal force against doors or each other.) He didn't say *before the turn?*... Before the turn. Oh, before the turn. O.K. good. What about yielding,

Traffic coming in on the right or the left. Yielding to traffic? In India, left side must, and right side must. Yield far left side, in India.

They must yield for left. (Note: Like the U.K. vehicles drive on *left* side of roadway.(For me, must stop. (40 second phone call delay.) Did the instructor have you back up the vehicle a distance? Ya. Good. (Taking my passenger through her home gate.) End. (Note: The main point here is that East Indian citizens who want to drive legally must undergo a minimum of *thirty* days of instruction from a qualified driving instructor.)

Interviews with Amy (Alabama)
Tim (NC) and Levi (OH)

Okay, we're jawing here with Amy on the way to 410 Union Street, downtown Nashville, on the 25th of May 2017. Tells me her daddy taught her, (in Alabama) some things. Not a whole hell of a lot but, one main thing stuck in her mind: Keep it between the mustard and the mayonnaise. For those who are not familiar with Alabamaneese, we're talking about the double yellow and the broken or white line. MAYONNAISE AND MUSTARD. And, then she got some schooling in high school in a simulator car, not the real thing. She definitely didn't take the car out on the Interstate, and definitely not during rush-hour. Is that right? (Nods, yes.) You received no professional training or paid for training from a driving school? (No.) So, when you are driving how would you rate yourself on a scale of 1 to 10 as a driver? ("A six," was Amy's answer.) Have you skidded a car on purpose, just to see how to handle it? When did you do that? It was raining on a little back road one night, just wanted to know if I could, so I did.

How did it turn out...when you skidded the car? I'm still here. Yeah, but what happened to the car?! Nothin' Oh, nothing? Straightened right up. Did you go in a circle? Yes. And, nobody else was around? Yep. (Proudly.)

Excellent. Have you been told how to handle a skid in the future? I think you're supposed to—if you're going left you hit the wheel to the right. Something like that. Opposite...(Note: Amy thought she should turn in the opposite direction of the skid.) (Corrected that notion.) And, you take your foot off the gas at the same time. Gotcha. Okay, so that's something she didn't know—or she had skewered in her brain. I've had other drivers say the same thing by the way in similar interviews. Okay, now, when you go on the Interstate here in Nashville or in Tennessee, tell me how that goes—what do you do? I don't drive here at all. Very clever woman—I wouldn't either. Why would I when I can get dropped off at the front door (Laughs). Don't have to pay for parking. Yeah, parking (expense) can kill ya. Okay, so that's the extent of your education, simulator and Dad, on the back roads of Alabama. O.K. Cool, thank you.

We're here with Tim, at the International Airport, Nashville. And, he's originally from North Carolina, or was in N.C. and moved to Tulsa, OK. And, was raised in Springfield, Texas, and that is where his dad. Well, first of all he was go-carting and riding motorcycles—what type of motorcycle did you ride? Suzuki 80. So, it was almost like a moped—kinda small, right? Yeah, small. Kinda easy? It was a pretty mean dirt bike for a small. Oh yeah, for a dirt bike. Oh yeah, they can be mean. So, it was a dirt bike, not a moped type thing. So, he learned how to control *a vehicle* at the age of 12 or 13. Which is great. Dad took Tim on a trip to...from Tulsa to...from Springfield Texas...to the Carlsbad Caverns, 200 miles approximately. Mexico? Mexico. After that, did you get in school, did you get driver education in high school? I did, I did in Homestead, Florida. My driver's education was my wrestling coach. I was able to drive just fine on the road. We got our miles in, but it was pretty lay-back. Did he take you on any of the major U.S. highways or Interstates? He did, with little tips along the way. Can you tell me about the little tips and pointers? Uh...It is hard for me to remember specifics, it's been so long ago. (Note: This is where I use a skill, developed as an interviewer to help persons recall better—it works and is a lot of fun.)

First of all, do you have a picture of his face? I don't, no. Okay, so in your mind? Oh, yeah, a picture in my mind. O.K. great. Now, how many hours would you say he drove with you? All total, about two or three hours.

Really? Was there any other education after that, either professional or paid for, or let's say a friend who was a really classy driver? No. So, its father at 12/13, and this three hours or so from this instructor in high school? Yeah. Interesting....End.

Levi:

Headed to The Turnip Truck (a whole foods type of grocery market), to have a holistic breakfast. He's from Ohio. My first question is: Where did you first learn how to drive a car or truck? I first learned from mother in Ohio. What age? (Here the recording skips) On surface streets, in a parking lot, on an Interstate or rural road? Yeah, I had to wait until I got my permit. Mom was sitting next to you? How many hours would you say your mom put in? Many. (There's a break in the recording at this point [unknown cause] continued with:) If you had to put a number on the hours behind a wheel with an instructor, what would that number come out? Gosh, my memory is a little foggy...I know it hit state guidelines. Off the top of my head I'd say 50. Let's talk state guidelines—when you were growing up and learning how to drive—what are we talking about there? I don't know what they are either. O.K, so you had fifty with a professional instructor, dual controls? Ah, no. Just sitting next to you, like mom? Hmm, hm. Did you go out on the Interstate—how many times? Again, it's fifteen years, I don't remember off the top of my head, I remember being on the Interstate. Ever in the rain or snow? I don't remember if it rained, I don't remember either one, but I would say there's a better chance of rain than snow due to the time of season that I was in. Can you give an idea of what the instructor stressed in the instruction period? (Thinks.) We went over everything. So, went over...(not coming to him). What salient points did he go over that you can remember? (Slowly thinking...) Put your hands...Like the old ten and two (positioning)? Yeah—just what to do and the red light, turn you know, and...I mean we just went through different, every scenario you'd want to teach. Again, I'm not giving you good data. Well I'll ask some questions to see if something pops up—what about following the car ahead—distance and speed? Yeah, we went over that. Do you recall anything about that, that you use today? No. How about handling merge signs and merging onto the Interstate or getting off the Interstate? I know we went over it. I don't remember anything that's valid. Okay. You drive now, right? Yes. And, what type of vehicle? I drive a very old SUV. <u>Okay, how old is *very old?*</u> Mid-nineties. <u>O.K and how often do you do that?</u> I live about a tenth of a mile from where I work. So, I walk or bike to work most days. But, you know, if I ever—I might drive a car a couple times a week. <u>Do you ever rent a car and take a long trip? (Boy is this showing interest or what?)</u> No, not really. <u>Okay, well great. Thank you for your comments.</u> Absolutely. End.

Interview with Rajat (India) and Ryan (Tennessee)

Rajat:

(From memory, not a recording) It's the morning of June 2nd, and I've just dropped off Rajat. He's from Nepal, and he's already given up driving here in Nashville, or at least in the United States for a while because he had an accident. He doesn't even know how it happened. Wasn't rain. Wasn't on the Interstate. Wasn't at nighttime. So, I quizzed him about, we only had a little bit of time. I quizzed him about what it takes to get a license legally in Nepal. He said you take a test of about 20 questions and you drive on a pre-determined route, doing left turns, right turns, using signals, backing up and so forth. That's about it. So, I asked him, who taught him. He says, well it was another newbie. Another beginner driver, which confirms my suspicions about the trouble here in Nashville (and other cities), is that they are under-trained. End.

Ryan: (Not recorded.) When and where did you learn to drive? I'll tell you something if it's not recorded. (Ryan wanted to go off record) Ryan's mother when he was age 14 took him out a few times in a parking lot, and, had him drive around. Then he got Driver's Ed. in high school.

He thinks maybe 8 hours on the road...but this was with *Brentwood Training Program*. (See "Phil" an instructor's interview at beginning of book.) Schooled in Knoxville, TN. (He's an accounting major from UT.) The school (High school is understood here.) did take him out on the Interstate. He didn't specify whether it was or not in rush hour. He did say, one of the last things they would have him do is take him out on the Interstate, otherwise they wouldn't pass, because they (Brentwood students need skills before they go out onto a major highway, like an Interstate which requires a little more confidence. His driver took him out on Interstate but not during rush hour. That's the *last* test, since students would fail without prior experience driving in general, Ryan disclosed.)

Interviews with John (KS) and
Thomas (TX), Plus Elaine, Chris, and Rajiv

John:

(Tells us how he got his start driving in Kansas—now residing in Nashville, in Dad's countryside pickup truck at age 14.) I grew up in the country in Kansas far away from pretty much everything. Driving my dad's pickup truck allowed me to go to school or work. <u>Okay, this was a standard shift, on the column?</u> It was "Three on the tree" (A saying I've never heard) <u>And, did you teach yourself most of time, or did your father spend time with you?</u> My dad spent most of the time with me and even when I would sit on his lap he would let me steer the car while he gave it the gas. I also worked a lot mowing lawns and pastures, driving tractors, and things like that. Internal combustion most of my life. <u>And, as you got into the later part of your school life, high school, was there an education program where you went out with an instructor?</u> I did like, I think it was a standard driving thing to get my license. <u>What did that consist of?</u> Driving with a person, like an instructor, kinda standard what a driving experience would be. How to parallel park, all your signs, braking, turn signals. Not performance driven though. <u>What type of hours do you think were involved with that instruction?</u> Some while ago, if I can remember, something like they would pick me up at school. They would show up after my high school was done and I would go for maybe an hour or two, couple times a week, for about for about a month. So, total hours about 10? I say 10-20 hours.

Any other instruction, either professional or non? None. Felt pretty confident behind the wheel. Okay, great. And, you've driven outside of Tennessee, right? I have, yes. Have you gone on any trip longer than 100 miles? Yes, many times. And in your vehicle? Huh, huh. How would you rate yourself on a 1-10 scale as a driver? Like as far as safety...Confidence as a driver. I would say an "8-10". When getting on the Interstate during rush hour, which you have done, right? Yes. Describe to me how you do it.

Depending on how much traffic there is, if there's a lot of traffic, try not to be close to anybody (John said "everybody"), because it will bottleneck... (garbled words) a lot of times. Make sure you have space between you and other drivers. You kinda have to just eyeball and watch the traffic that is already on the highway. Who else is in front of me, and wait for a gap to come in. If you see an opening, have to adjust your speed, try to match theirs, signal them and merge.

O.K. John, so you look for an opening and you try and match the speed and you merge? Correct. <u>Good. How would you define the word merge?</u>

How would I define it? Well, in a car we don't have verbal communication with other drivers, so we use things like turn signals. We can use our hands to wave, make eye contact with somebody, it's kinda different in every situation...but it's just one of those things that just kinda happens. Both drivers that are involved in the merge are paying attention. Sometimes they're not. It's also judging to if you are merging in, you have to think about how fast the other person (driver) is going, and what the performance limitations are of your car. Should I speed up to get in front of him/her or are they going too fast. Should I wait and fall in behind, that kinda thing. <u>Notice: Up until now the definition of *merge* he has not simply stated.</u> (This *lag in answering* is a sign that the word and its meaning are not well understood, and so poses a hazard to and by drivers who have never defined the word, yet encounter it almost daily. In high speed travel, any traffic flow, this is unacceptable.) <u>How would</u>

you define the word *merge*? The coming together of two things. In the sense of driving it's two cars coming together in the same lane. (Unfortunately, that definition would also apply to two vehicles colliding with each other —you see my meaning?) And, **then**, John qualifies his definition thus: But they can't occupy the same space. How do you take turns at high speed—let's say you are going south on I-65 (a North / South highway) and you see, ah 440 (East/West). I want to go over to Green Hills. It says (speed limit) 45 mph. Or 21st (Ave.) Yeah. It's also depended on traffic, that turn. I know is a very kinda wide gradual turn. My particular car is very low to the ground, so it has a very good center of weight (the term is *center of gravity*). The suspension is also rather tight. So, I know on those kind of turns, because I know the feeling of my car, I don't have to slow down a ton. Now, if I were driving an SUV that has a higher center of weight, something like that, kinda feel it roll a little bit. At what point to you begin to slow down? (This question asked of another driver I interviewed earlier.) I'm usually coming into a corner like that will let off the gas just before I even start to take the turn. Start to coast. And then maybe tap on [my] brakes just to let the person behind you know that you are slowing down. And, if you just let off the gas they might not recognize you are slowing down? So, you can signal them with your brakes and your turn signal. Once I'm into the turn kinda judgmentally judge what it is going to look like, I will slowly start putting on the accelerator again and push through the turn if I'm going a little too fast. End.

*Not necessary at all. It gives a false signal to drivers behind who should also be trained and aware that the slowdown occurs well before the turn.

Thomas: (For brevity, several of Thomas's answers are given in short. He's from Texas. Parents funded driver training.) How many hours on the road with instructor? I'd say in Texas. High school? No, just part of getting license go to driving school. How many hours? 12 to 20 hours. Drove Interstate? Yeah, I did. Did you get a chance to drive in the rain or rush hour? We drove in heavy traffic. (Rates himself on scale 1 to 10 a "7" as a driver. Drives a small SUV.) Do you ever go through the gears to slow the vehicle down? Not in town. You understand the fact that the gears speed up and slow down the car—you observed that yourself or did your teacher teach you that? My instructor taught me that.

<u>What is the best time to shift to the lower gear?</u>[3] Before you go down a hill. (True. Also, upon slopes upward and inclines it is safe to manually or using sequential gears go to higher gears since the slope automatically slows the vehicle so that a smooth transition and more slowing is obtained. Same for flat roads and highways.)

<u>Have you seen when you are driving, have you seen drivers who stop at yield signs unnecessarily.</u> Not sure, maybe.

<u>When you are on any Interstate how do you maintain distance from your vehicle to the vehicle in front—what do you use as a guide?</u> I probably just use experience now, but, when towing a trailer, I use more, if not I use less, as I was taught to do. One car length per second (Use what? On this specified question responding with vague answer is not a good sign that he has been trained properly—then he corrects himself): Sorry, one car length per 10 miles per hour. <u>That was the standard. Then it became two seconds. Like if you [vehicle in front] passes a pole, or some kind of marker—you pass it two seconds or later that was good. Now they have upped it to three.</u>

3- Lower gear here means turning *slower* because it is a larger (diameter) gear, not lower in number.

Which shows the deterioration in perception of our driving community. If that is not an indictment, I don't what is. What's your opinion of these signs, these neon or traffic notification signs that keep saying how many people are dying on our Interstates or highways? I think some of that is good. I don't love the *cheeky* ones...they make driving seem trivial. There's a couple of good ones, but most are flat, don't have much meaning or punch. I wish there were more positive—like the "Please keep Nashville litter free, instead of "Don't Litter/$100 Fine. The former is an appeal to the positive side of life. Instead of always telling people "Don't do this. Don't do that." People get so, it's a hard response to the word "Don't" that after a while it becomes ineffective altogether. End.

(Driving to taco restaurant with Elaine on Thursday: she tells me that there is 200 hours of driver training, behind the wheel in the state of New Hampshire plus theory.) (Note: New Hampshire has the lowest traffic fatality rate per capita of any of the 50 states.) Continue. Then actual course work, you had three or four months. You had classroom hours then additionally to classroom hours you had more behind the wheel driving time with the instructor. <u>In your 200 hours, were you taught how to yield properly?</u> Yes. <u>Taught not to follow too closely?</u> Yes. <u>Were you given a chance to skid the car, let's say in a parking lot, where it wouldn't hurt other people just to handle a skid?</u> They had us drive through a parking lot of snow to get what it felt like (garbled). <u>Fantastic. None of what you told me is done here in Nashville.</u> <u>If it snows they don't take the students out, if it's raining they don't take the students out on the Interstate or even out on the highways.</u> How are they going to learn? <u>That's why the book is being written.</u> <u>So, how many hours of classroom?</u> Three months, and one class two or three times a week. Then separate from those classroom days, that's when you would set up schedule days to go out and drive thirty to forty-five-minute period. You had to have a certain number of those scheduled with an instructor.

Chris:

(It's Sunday, 28 May 2017)

(He just told me he had a fender bender on Granny White at the corner of Tyne Blvd, which is a four-way stop, and it was raining yesterday. The car hydroplaned down from 35/40 mph and hit another car in the intersection.) Where did you first learn to drive? Knoxville, TN, rural roads at age 15. First license? 16. Did you have any formal training in high school or otherwise? Yeah, I took a driving class. And, how long were you behind the wheel in that driving class? (At first not sure, said 4 hours) Less than twenty hours. Did you have an instructor on the passenger side with dual controls? Yes. What were some of the main things he tried to instill in you as a driver, newly? The usual things, proper speed, that kind of stuff. Did you go out on the Interstate several times? I think so. Interstate 40? Yes. Did he have you back the car up any great distance? We did parallel parking. (No backing up.) How about going out in the rain or rush hour—not that Knoxville has much of a rush hour compared to us/Nashville? We were in traffic, I don't think rush hour. Moderate traffic. How do you handle yield signs, going on the Interstate—tell me about the speed of your car? Actually, merging onto the Interstate, I would be aiming to be up to Interstate speed. Do you ever slow down or stop for the left-hand lane? No. (That was all the time we had for the interview in the car.)

Rajiv:

(Started with dad, 1998. No formal instruction after that.) Start with a standard shift or automatic? (Says it wasn't a new car so standard. Standard but now drives an automatic. Not a sport automatic.) (There's a lot of road noise on this recording making it difficult to transcribe exactly his words.)

How about in high school, were you given any training or classes? Yeah, I did take driver's ed. What did that consist of? We had a driver's ed. car, dual controls. Where and when would they take you? We'd drive around the city. No interstate practice, or in the rain/snow? Teach you how to skid a car? No. Did they teach you merging and getting off properly from a highway? No...

I'm discovering that there are a lot of under-educated drivers in this country, with hand-me-downs from dad and uncle Joe. (Rajiv agrees.)

Interview with Trevor (Buffalo, NY)

Trevor:

(Driving with Trevor from Buffalo, New York. Recently moved down to Vanderbilt area - Nashville.) Where and when, and who taught you how to drive? "The first time I drove a car was on a Saturday morning right after I had gotten my permit at the DMV in Buffalo. My dad gave me the keys...baptism by fire, I was 16 years old. He handed me the keys...sure enough I didn't know that as soon as you take the car out of park, put it in drive, if you don't have your foot on the gas it starts moving a little bit. That kind of sent me through a loop. On the 15-minute drive home I was heavy on the brake. First time I drove a car and learned from experience. When did you first attempt to get on an Interstate? "Probably two or three months after that. Was that I-95 or 90? What difficulties do you sometimes encounter during Interstate rush hour traffic? Coming from a smaller city, just volume, especially the Interstate I'm accustomed to, the stretch on 90, in Western New York is only two lanes, east and west bound. If a truck ever needs to pass, that can get a little dicey because there are only two lanes. What is your reaction—what do you do, your mental process? Generally, I'll try and pass trucks before they pass me.

<u>When there are no trucks, how do you judge the distance between you and the car ahead?</u> In New York we have a set of mile markers in the middle of the Interstate. (Garbled words.) They have 3 posts between every tenth mile. So generally, the rule of thumb, you want to say…one car length per second. (He meant per 10 mph.) (Also note his answer came *hesitantly* – and that as stated earlier is not a good indicator of skill or adequate for reaction time on highways.) <u>That's the original one. The new one they came out with several years ago: you pick out a post or a sign and you count like that red truck, (author indicating) one thousand one, one thousand two, one thousand three. And, expect to be three of those (3 seconds) between you and the car ahead.</u> But that's good, one car length per 10 mph is good. (Certainly, much more than is allowed on Interstates that I have traveled.) (Learned on an automatic with Dad—same type transmission now. Drives a Honda CRV.) <u>Do you use gears to slow the down vehicle?</u> If it's heavy rain, I'll put it in low gear. <u>How about city traffic?</u> We don't have much city traffic up there, so generally don't have to. <u>When it comes to merging onto the Interstate 90?</u> New York has toll booths on every entrance and exit.

Obviously, there's the branches like 190, 290, 390... <u>There's no just clear getting on ramp?</u> There is generally, it's just a longer third lane comes off the toll plaza. <u>For getting off the highway?</u> You try to get up to speed, with everybody else and then shift over. The general rule of thumb is if you see someone trying to merge on, obviously if it is safe for you to move into the left and you're in the right lane. Just move over for him.

<u>What about emergency vehicles, do you have a law like we have here?</u> Yeah, State troopers and emergency vehicles: the law states you have to move over if safe. <u>Now, if there are no flashing lights on an emergency vehicle, what do you do?</u> If I see it parked on the shoulder and it's safe to move over I'll generally always move over, regardless if I can see the maintenance worker, or whoever it might be. <u>Speaking of moving over, describe to me how you move from one lane to the left, the faster lane?</u> Generally, I check the review mirror, check the side mirror, over the shoulder, if that looks good I move over. <u>What about signaling?</u> Oh yeah, of course signal before you check the mirror. <u>How long do you do that before you move over?</u> Generally, try to give it two clicks.

<u>What happens when you see an accident, emergency vehicles, a collision?</u> Tough to say, I haven't stumbled onto one. On a two lane, generally there's a lineup, I'm stopped before I see one. <u>Stopped?</u> I'll be in a line of cars. There will a be a line up or a slowdown before I reach that point.

<u>You're not a doctor or a medic-so there's nothing you can do about the injuries or collision, right?</u> If I were a witness to the accident, it happened immediately, I'd try to pass the scene of it, pull over where it is safe and try to help somebody out, and if it seemed I was the first one to stumble upon the scene. <u>What if you were the tenth car?</u> And no one had pulled over? Probably pull over. (Note: Trevor qualified my question to him. At this point I had no more questions that were <u>not</u> oriented to Nashville driving conditions.) Sure, you will. You definitely care about this stuff. End.

Interview with Geeta and Glen

Geeta born in America, (Father born in India) <u>From whom did you learn to drive?</u> My brother in Texas at age 15. <u>Was this on a rural road, an Interstate, or a parking lot?</u> In an urban city on the side streets of our neighborhood. <u>With a permit?</u> Yes. <u>And what do you recall he mainly tried to stress in his instruction?</u> Safety #1, following the rules, and generally getting comfortable. Being a proactive driver. <u>What about speed?</u> Yeah, we talked about speed. <u>Well, what did he tell you to do about speed?</u> You know I don't recall anything specific about speed. <u>Would you say that you were always within the speed limit of the street that you were on?</u> Probably not. <u>You'd go with whatever the flow was, that pretty much is the covenant.</u>

<u>After that did you have any formal training after that?</u> I did. You have to go... <u>How many hours behind the wheel with an instructor?</u> I would say I did the course over a week. So, maybe three hours behind the wheel. <u>In that three hours did you go out on an Interstate highway?</u> No. <u>Did you drive in the rain?</u> Probably not. <u>Did you back the car up under supervision?</u> I did, yes. I definitely had to park and reverse back up, yeah. <u>So, what happens when you get on an Interstate now?</u> I only drive on the

weekend. It was in Wash. D.C. <u>I-95 runs through there, did you get on the 95?</u> No. I go on 395 or Route 66 which are the two bigger highways a lot. <u>How do you handle the fast-moving traffic in the left lane as you come up on the merging lane?</u> I only view to try to make sure I merge in appropriate spaces in the flow of traffic." <u>What if there is a tractor-trailer on your left, doing about 65/70 (mph)?</u> I sort of try to avoid large vehicles like that. <u>Tell me about that, how do you avoid them?</u> "I'll either try to go in front of them or the next lane over but directly not behind them. <u>Have you ever stopped and waited on the shoulder 'til the truck passed?</u> No, I think that would be very dangerous. End.

(Driving with Glen.) <u>Where and when did you first learn how to drive?</u> In Dallas, TX. I got my first permit when I was 15 and received my full license when I was 16. <u>Do you drive daily?</u> Yes. <u>What difficulties do you sometimes encounter during Interstate rush hour traffic?</u> People here, especially on this stretch we're about to get on, from Almaville Road to Hayward Lane, don't recognize the geographic anomalies in that stretch of road. Just the hills and the twists and turns, when people don't realize what lane they need to be in. Let's call them drivers. So, drivers aren't aware of terrain and rain, even though they've done this drive for the past five to ten years. They tend to put it on auto-pilot and react instead of plan. I've observed that. Give me an example of, recently when you've seen something like that, so we have something concrete for the book. So, the driver yesterday-I drive pretty defensively, I watch out to see what another driver is going to do, and in really thick traffic I saw this guy in a red pickup cut across three lanes to get his exit. Really wasn't considering the whole picture? Yeah. Good, that's what I wanted, an example. I've seen this happen many times. In fact, two days ago I decided to put that into the book as well. You know, the changing of lanes after one click of the signal, or ½ click in heavy traffic of the turn (lane changing) signal and move over to another lane. Especially, a faster lane. Yeah! OK, let's go back to your beginnings, in Dallas. You got your permit what training did you get and from who? The school district put up portable trailers and we did driving simulations-through the summer—two or three hours a day. Like in a parking lot? Yeah, in the parking lot in front of the high school. Me and nine other kids with this driving simulator that was really outdated from the '70's. But, it put you in situations, grade you on your reactions. Almost, like a video game. Yeah, Yeah. How many hours do you think total? If it were a three-hour course Monday through Friday, probably did thirty hours of classroom in two weeks. Okay, what about in the simulator? That was the simulator. The rest was theory about traffic signs and lights and lanes and all that? And drunk driving and things like that. Now since that time, and before you got your full license, did you have any professional or other training from your parents or siblings? So, my dad and my brother-in-law: both took me out in their cars and let me practice driving on city streets. What type of traffic? Usually, after dinner so it would be a little lighter than day. Was the sun still out or was it night? With the sun starting to set. And then, practiced parallel parking, an element of the test. After the two weeks of classroom—you know I want to say it was one week in the classroom and one in the car. That's what it was. We're going back to the driver's ed. in high school? Yeah. I'll correct that, that's fine. We would get in the car where the instructor would sit where the passenger would sit. He would have his own brake pedal. Oh, so it went from simulator to real car? Yeah <u>Oh, that's important.</u> Yeah. Classroom slash simulator then car. <u>But still under the control of the same environment, this tarmac--is flat area?</u> We actually drove on highways. <u>No Interstates, no I-90's, I-95's?</u> I-65 in Dallas. <u>But, not in rush hour?</u> Not in rush hour. <u>And not in the rain?</u> And not in the rain. So, we practiced with the instructor behind the wheel. So, then brother-in-law, father, learning how to shift the manual transmission, parallel parking, is part of the exam with the state trooper. <u>How about backing up the vehicle a distance?</u> Don't think I practiced that very much. All right. So, <u>let's ask some poignant penetrating questions about you and the road when you're driving your Jeep Wrangler. What do you observe, and what do you use to guide your vehicle as far as the vehicle in front of you on the Interstate?</u> I do it by feel. I've been driving since, for 32 years. I like to be a safe distance behind the car in front of me. <u>Tell me exactly about that, how you judge that: like an instance like this...I point to car ahead.</u> This is a pretty good distance for me. Probably this is four car lengths? <u>You know why I'm doing this? Because there is going to be slow-ups all along in this section here until after Thompson Lane.</u> (Traveling on surface streets.) A part of the reason I do it in this lane is because drivers are going to want to merge over. Yeah, I also want to let drivers merge over and not have crunch time. So good, what rule of thumb that you've been taught do you use? There's a thing about every ten miles an hour you count a second. Right, you still use that? I do it when I'm driving long distances, when I'm on cruise-control on the highway. Yeah, what about in this situation? This situation I just do it by feel. OK, give me an example the last time, yesterday or the day before—obviously the traffic was no different. What would you say, let's say you are driving this car now, how close would you get to that car in front of the truck? I would get no closer than where that blue van is. Blue van, blue van-oh, I see, blue van, if we moved it over. At this speed you would allow at least one car length? Definitely. OK, good. All right. What, and this is good, this is a good question because you are old enough to learn on a standard shift, the standard-shift your father was driving was a three-gear-Low, Medium, High, no six forward gears or five forward gears? Yeah. This may not apply, how about braking instead of gearing down to slow the vehicle? I learned how to gear down because my Dad says, He didn't want to pay to replace the brakes, although I guess replacing a transmission would be more expensive. (At this point the author did not interrupt Glen to note that using the vehicles transmission to gear down, if done properly, *does not wear out* the transmission.) <u>What did father teach you about how to do that correctly?</u> You know, man, it's been a long, long time...<u>Because you don't drive a standard, do you?</u> No, it's been a long time. <u>OK, just take a moment, because there is a way with a 3-speed, low-medium-high, there is a way to do it properly.</u> Probably, if you're just, like coming to a stop sign I would let the speed taper off, of its own volition, and then shift accordingly. <u>Right, you would not do it on a downhill? You do not shift down to a lower[4] gear on a downhill. It's only done on an uphill.</u> Yeah. <u>That's empirical evidence I have already observed and put into the old vault of information on driving. Good! So, you came up with the right answer. Now, on a downhill you would go from Medium to High, or Low to Medium. There would be no real liability doing that.</u> Right. <u>You want to increase the speed or momentum of the car when you throw it into a higher *numbered* gear, you need more forward motion because the gears (that</u>

4-Again, "lower" translates to larger slower turning gear. (Usually gears number 1 and 2 or (S")) are working) are now smaller and they need more force to drive them. Like, when you ride a multi-speed bicycle, you know how when you go to a lower gear your feet start doing this. (Demonstrating more push.) Right. Yeah. Good. All right, what have you observed about cars on the Interstate merging that stop on the merging lane for traffic to the left? Oh yeah, that's one of my-I've got many driving pet-peeves, but that is one of them. You don't stop on a lane, especially if...Cars behind...Yeah, if you have cars behind you, or you just take a risk and merge. Yeah, there's a little bit of a risk and the worst that can happen is you pull over into this lane here-the margin. (Author indicates the shoulder) It's really crucial. People just get terrified and kind of camp out at the end of the merge lane. OK, so fourth question: Also merging onto the Interstate more slowly than the traffic traveling in the immediate left lane—have you seen that? You see they come, kinda of drift over into the left lane, here in Nashville—if I were doing 60, 65 mph and they kind of drift over at 45 and 50 mph? Yeah, and they're not paying attention to the speed of the traffic that they are flowing into. Yeah, good, OK. Now, what about changing lanes, well we just talked about this, changing lanes abruptly in heavy traffic, that's done. What do you do when you see flashing lights in an emergency vehicle on the right shoulder? You get at least one lane over. All right, that's the law... The other part is slow down to under 45 mph. But if it's *safe*, you are not forced to move over into this lane if they're on the shoulder, if it is safe, you don't do the thing we just talked about. You don't just pull over in front of another driver. I've seen that happen. What if there is an emergency vehicle stopped on the shoulder, the lights of the emergency vehicle aren't flashing? Any time there is a car in the right shoulder, I either change lanes or slow down to 45. If there is no flashing, you will attempt to get into the other lane? Yeah, because there is going to be a person present. OK, good enough. (The law in June of 2017 in Tennessee states exactly that, any type of vehicle that is in the margin.) Guess, we call that prudent driving. All right. When you see an accident coming up, there's been a collision, there's merging vehicles, there's police cars, what do you do—regardless of the amount of traffic—whether there's no traffic or whether there is very little traffic, what do you do? There's an accident, there is obviously some kind of kerfuffle...Yeah. on this side. I'm usually trying to get into the farthest lane away. In case there is an explosion or something. (Chuckles) Right, oh really? Yeah. All right. What if there is a slow-down, drivers are creeping by it, because they have some lurid interest in what's going on? I hate that. You have observed that many times, right? Of course, yeah. There's a weakness there. All right, left hand turners that turn too early into the oncoming traffic, not allowing for cars ahead (in opposite direction) to clear the intersection, have you seen that on surface streets? In other words, turning too soon[5]. Yeah, people not cognizant of the fact that there are oncoming cars. Drivers in Nashville generally are not cognizant, whether it is landscape or other drivers. It is pretty frustrating. Now, where do you look, where is your attention focused on the Interstate when you are driving —I know they say all around, on both sides and behind Keep the attention spanning. Where do you look 75% of the time? I focus several car lengths. As a rule? Yeah. As a rule—that's a good thing (practice). On a scale from 1 – 10, how would you rate your driving ability? A nine. Good.

5-Happened to author in 2017 in Nashville city proper at the corner of 4th Ave and Koreans Vets Blvd, forcing him to jam on brakes with car full of chauffeured passengers, ruining (warping) the front rotors—but no one was even discomforted.

Interview with Carlos (Philippines)

(Driving Carlos to BNA (International airport) from Franklin, TN. Carlos is from the Philippines, his dad taught him at age 12!) <u>What were the requirements when you were growing up to get a license?</u> Requirements were you could drive at age 15, with school four to five hours a day. <u>How many days?</u> Five hours a day back in the day, my youth. <u>I'll fact check that on the Internet.</u> (For the present requirements.) Five days a week. (Inaudible but recalled by interviewer.) You have to complete the course and then do like a test for learner's permit. <u>What did the test consist of?</u> Road signs, rules and all that. About 50 items. <u>What about driving?</u> You have to pass first, complete the test and then all the passes including the driving. <u>How many of the hours, five hours a day, 20 days are devoted to *behind the wheel*?</u> 10 sessions of driving, takes one half hour each time. <u>In a simulator or actual car on street?</u> Actual car. <u>Do you ever go onto the major highways?</u> Yeah, you have to go first on the streets. Back then we don't honor the Stop. We don't have any Stop signs. The red stop-sign you have we don't have. (Really?!! I interject); Yeah! <u>Yeah! Is that still the case?</u> Not anymore. We have to really watch where we're going, a lot of people walking. I guess it's more a discipline...we're honoring now. You really need to be disciplined. Imagine four streets, four stops but there are no stop signs. <u>So now, you are on the streets in a car with the instructor here...do you have dual controls?</u> No. <u>Can the instructor stop the car?</u> No. <u>Can he steer the car?</u> No, just me, just you alone. And it is a manual (transmission car). <u>So, you're out for about a half an hour to an hour with this instructor who doesn't have control of car if you lose control?</u> " Carlos responds with a Hmm Hmm (Yes).

Then you are on city streets without stop signs. You have traffic lights? Yes. <u>And it's not in the rain, he doesn't take you out in the rain, does he?</u> No. <u>Does he (did he) take you out in rush hour?</u> Yes, twice. <u>Only twice, so, maybe it was only for about an hour?</u> Yeah. It's hard. But you are still learning. And, if you stop sudden they give you the horn. <u>Right, so you're nervous, you don't want to mess it up?</u> It's a manual transmission so sometimes you hit the wrong gear and you stop. You are panicking. <u>How many hours did you dad work with you in the car?</u> "Every day. Every single day. <u>What did your dad teach you?</u> Mostly be patient and don't get hot-tempered. If you have some problems in the street and some person at your back blows his horn don't get rattled and mind your own business. What about taking turns at speed? Yeah, always observe your speed limit, always signal, left right. The textbook. In our DMV version you really have to follow the rules, or they will not pass you." You have 3 and 4 lane highways? Yes, similar to this I 65. Did father train you over and over again how to merge correctly onto the fast-moving-highway? After I got my student permit. Yeah, 'cause you can go there at 12 years old drive on the freeway if you have a professional licensed driver. Some do that, but my dad is from the military. So, he would take you on the heavily trafficked highways and have you get on and get off? Yes, when I got my learner's permit. What did he teach you about following the car ahead of you? Just mind your distance, I think 3 car lengths, you know, apart. Mind your distance apart, your speed, obviously all the time. What kind of speed limits do you remember you had in those days? Then? To tell you honest there also were no speed limits. Really? We're not following the limits...but there's one hundred percent, kilometers per hour so, like 60 on the streets...freeway, we call it highway, 100 kilometers? Maximum, yeah, I think a hundred. What difficulties did you encounter in your first year in rush hour traffic or heavy traffic, did you have any difficulties? Yeah, the most trouble is the traffic and congestion, but there's a lot of cars on the road, and most of them are undisciplined, unskilled they just swerve where ever they want. We don't have we call it...officer? Back then that much. Late 80's. Did you ever have a collision? Never. I'm always getting accidents here...getting my rear ended. Oh yeah? Yeah, mostly stopping, like you know freeway? Fortunately, not so hard so, but too close to you. Yeah, following too close-I find that's one of the major weaknesses. That's why I keep my distances. Yeah, look what I'm doing-ten car lengths behind vehicle in front. And my son, who is 14 now always tailgates. I tell him: When it's time to do your driving keep the distance that I have. Don't follow too close. You never know when the guy in front of you stops. Yeah, either you'll hit him, or someone will hit you. End.

Interview with Allison (North New Jersey)
Marissa (Chicago)

Allison:

(We're traveling to the Johnny Cash Museum on the tenth of June '17 with Allison from North New Jersey)
Who was the first person who helped you to learn how to drive? My cousin's girlfriend. (Erin) Your age at the time? Sixteen, in my town and she brought me to my middle school. And, we learned how to drive, and how a car behaves in a parking lot. This is before you had a permit, right? Yes. Now, after that did you get any formal training in high school or from a professional driver? I had to do three or two classes for six hours with a training driver. Was it a driver training school? Yeah. A driver training service. What did that person help you to learn? (Smiling / cheerfully spoken) Well, the first thing they did is take me out to the highway in Hoboken (N.J.) So that was gotten right in. Good old Hoboken, what highway is that? 80...During rush hour? No. And, not during the rain, right? No.
What are some of the [points] that you can recall? Blinkers (signals), braking...What about the flow of traffic merging? Yeah, merging (she's recalling) How often did he teach you to go on and off? Don't remember (still cheerful) You say six hours behind the wheel? Yeah, we

did two, two and two (hours) <u>Excellent, what about the car you are following ahead of you – give you a rule to work with?</u> A car or two behind. (Note: this is insufficient gap for any normal highway scenario.) <u>Did he attach any significance to your speed?</u> (Lag here in replying) (So, I coached a bit.) <u>Because the old rule is one car length for every 10 mph.</u> Yeah, I'm sure that was it. (Laughs) <u>That was it—I'm helping your recall. That's fine. (Allison is laughing.) And, what about taking turns at speeds—like we're coming up to a fairly sharp turn?</u> To slow down...<u>When did he tell you to do that?</u> When I was turning because I didn't listen. I drove with my mom after that I sped up around a turn (a few garbled words here.) <u>You know, it's O.K. to speed up once you are in the turn?</u> (She acknowledges knowing that.) <u>Okay, tell me about these questions—I've got a couple of canned questions for Tennessee drivers. What difficulties do you sometimes encounter during Interstate rush-hour traffic?</u> More traffic--not happy, especially New Jersey. You've got the hands and the swearing (undecipherable). <u>Well, that's the influence of New Yorker, probably.</u> Going on the shoulder...<u>Really, to get around?</u> Yeah. <u>Tell me about [drivers] changing from the right lane to the faster left</u>

lanes—an incident if you recall it. (She mentions the habit of putting on directional signals but not waiting before changing but just changing lanes.) During rush hour? I have found that here a bit in Nashville. They put on their blinker *as they are changing lanes.* Or they don't do that at all. What about following the vehicle too closely in front, do you see that in New Jersey? Yeah. (Author acknowledges and that it happens a lot here in Nashville.) What about SUVs, following behind these [wide] SUVs in your Jetta? Yeah... Kind a hard to see (much ahead), Yeah, you don't know how the traffic is going. Now, how have you normally kept your attention on the road ahead —where do you keep it mainly? I like looking in the rearview mirror a lot. And, a lot!? Why is that—you are heading forward—you think you might get rear-ended? YES.

    Did that happen? Yeah once, which is probably why. All right, let's go back *before* the rear-end. Where did you put your attention in terms of distance on the Interstate? (Allison indicates just the car ahead.) Now, you know the way to do it is to look five to ten cars ahead, so you can tell the flow of the traffic, right—you've been taught that haven't you? (Laughs. Apparently not.)

Of course, that's when the big SUVs get in the way and make it impossible to see. All right, number two: did you learn from your cousin's friend on an automatic transmission or a manual one? (Automatic-never has driven a manual transmission.) Even in your automatic (Jetta made by Volkswagen) you have, I'm pretty sure, sequential gears, tell me about that. It didn't really go well—it would just do it if I didn't do the right thing. Well, do you have any knowledge or instruction on how to slow a car down using the gears? (she answers no,) Therefore you don't do that? No. Yeah, most drivers don't know or don't do that.

How about have you ever seen a driver stop on a merging lane onto an Interstate? (She indicates that she has.) How often would you say you've seen that the number of times you've been on the Interstate. (She indicates several.)

How about too slowly onto the Interstate or slower than the traffic? (Again, she indicates yes.)

What do you do as a driver when you see an emergency vehicle? Pull over. (I had to further find out what she meant here.) Get over to the next lane.

When there is an accident on the *other side* of the Interstate, do you slow down for that? (Note her answer here.) Everybody does. They're so curious. Yeah, not a good thing. (Unless one enjoys more rear-ends, traffic tie ups and slows than usual.) What would you really want to do in that situation? Get going. That's right, move over and get out of that area.

(Note: When we *really think about it,* Interstates have no stop signs and no traffic lights. There should never be a slow-down to a point where the cars come to a stop. Come to a stop. Completely unnecessary.)

Have you seen left hand turners turn too soon into the oncoming lane?

(Again, she indicates this is something that she has observed.) (At this point Allison's destination is coming up so I end the recording and thank her for her candid replies and data.)

(Using sequential gears in any car or SUV going up a slope or hill is advisable. For one, downshifting increases control that is lost when simply braking, and in wet conditions this could prove pivotal to the stability of the vehicle and thus the safe transport of its occupants. Two, in conjunction with the slowing effect of the incline, gravity, and braking simultaneously can avoid collision with a large vehicle in front that has come to nearly a stop, such as a fully loaded tractor-trailer or fully loaded bus. Conclusion: the use of gears, sequential in an automatic or standard transmission also to slow a vehicle is not only for navigating down slopes.)

(Marisa from Chicago, visiting Nashville, born in upstate New York. She like a lot of drivers I've interviewed got started driving with her dad.) <u>Was it in a parking lot or on a street?</u> In a parking lot. <u>Did you start with a manual shift or an automatic?</u> Started with a manual. <u>You'd be surprised how many women have been taught by their fathers to start with a manual.</u>

<u>Back in Chicago, what do you notice about drivers following too closely to the driver ahead?</u> Feel that tailgating is not that big of a deal that much. In New York it's more prevalent. <u>You used a term tailgating, now that's a technical term in traffic regulations. It's when a driver's vehicle is practically within a yard (36 inches) of</u>

the vehicle in front. I'm talking about just following a car length or less than a car length behind at 60/70 mph. Have you seen that in Chicago? Well one, I would say rarely are you driving that fast...rarely not in traffic (garbled). I don't drive on the Interstate that much. That could be it. Let's scratch that question. Here's something you know about. You know that you can use your gears to speed up and to slow down the car, right? Yes. How often do you think you do that? Well, I don't drive a manual anymore, but when I did I would say that is probably...highway driving?.80% of my slowdowns. Oh, really?! Yeah. Really, 80%--driving on a non-rush-hour basis? Yeah. REALLY? My dad told me that's what you do. Yeah, you can! My dad was always about you shouldn't use your brakes on the Interstate. Oh, I should have met your dad! It's exactly my philosophy. Yeah. Did you know that at one- time race cars didn't have brakes? (She didn't know that.)

　　How often have you seen drivers stopping on the Interstate on-ramp merge lane instead of smoothly merging? Way too often. It should never happen. All right how about changing lanes abruptly? All the time. (Inaudible.) What about slowing down to view an accident, especially on the other side of the Interstate? That's how you get most traffic jams. Have you seen left hand turners turn too early into the oncoming traffic— this would be on surface streets? That one doesn't jump out at me. How about drivers at red light have their turn signals on (no cross traffic coming) and they wait and wait? Chicago, most streets are no turn on right, and streets that aren't that way tend to be pretty heavily populated with foot traffic. So, there's not a lot of opportunity to see someone turn on right. All right, going back to my first question: did you get any training from your dad more than when he took you around the parking lot—like from either high school instructors or professional instructors? No, just my parents. (Paused and then ended recording.)

Interview
Patrick (Rwanda)

We're in the car with Patrick, whose true home is Rwanda, East Africa, and he did get a license when he was in California about ten years ago. How does a citizen of Rwanda get a license to drive? The training is very tough. You must pass a written test, which is also tough. What are some of the things you are tested on as a driver behind the wheel with a tester or instructor? It's mostly the basic ones the ones that ...but the problem in Rwanda is you have easy roads, small roads, you have so many people working, so it is really rough. Do you have fast moving interstates, like we do? Unfortunately, no. But, that is one of the reasons it's really difficult to drive there. Sometimes we bump into people and so if you are taking driving license then you fail. So, in the testing you go on small roads, big roads and paved roads, obviously, But not anything major, no major 70 mph Interstate, okay. Oh, also roads where there are workers. How about in the rain, do they take you out in the rain—test on how you handle a slippery road? I try to do it. In Rwanda you have a long season of rain. So, it's not surprising that you might be taking the test during the rainy season. Now, is this the test or is this the instructing you have to pay for from a professional driver? You have to pay a professional driver to teach you. And now, it's prior to going to the school. Do you think this is now true, even though it's been ten years or more since you were there? I think it's true.

Good, I wanted to establish that so as not to give out stale dated or now false information—did your father or mother begin teaching you a little about driving when you were younger? No, I learned how to drive when I was older. When I was eighteen years old. It was an interesting phase. I was very cautious. I didn't have the teenage vibe when I was learning how to drive so as an adult you don't do that. You are very aware of your...So you didn't get a learner's permit like a lot of kids do at sixteen and seventeen? No, I didn't do that. What about-did any of your friends try and teach you who already had licenses who were maybe a little older? Yes, yes. Okay, how did that go? It went well, yeah. When I took my driving test I actually did have friends taking them too in a shopping center, big parking lot... You didn't get any training from a professional? That was in California. Let's go back to Rwanda—in Rwanda you had no training whatsoever. No. Okay, it's only when you got to California, and you already graduated high school or college? That is correct. And these were your friends who are also from Rwanda, or from America? From America. Ah, OK, they were in school with you. Now it's clear. So, they said, "Hey, here is Patrick, he's from another country, let's get him a license because he's a nice guy and we like him, and we want to go out with him and have fun. And Patrick said, "Hey, that's a great idea, how about teaching me in the parking lot. So, that's how that happened? That's how we (garbled). And after that, no professional training and no school training from college or from anything like that? No.

Then, when you went to get your license in California, which I have also done—I have had a CA license, how did that go? It went really well. As I said, I was very cautious. And what did they ask you to do in California? They take you on a commercial highway. Then they ask you to essentially see how you drive and merge. There are so many California highways you gotta learn how to merge. I took my driver's test on a rainy day, so it was testing my (garbled). They took you out on a rainy day? You know why that is—it's so rare to rain in California...(joke) (chuckles). I was so afraid that I was going faster. I was under the speed limit. Oh yeah, so you stayed under the speed limit and did really well. Did they ask you to back up a distance? Yeah. They don't ask you to do that here (in Nashville), backup 10 or 15 feet. All right, merge onto the highway, travel at least one exit, get off, backup, park? Yeah. In the rain? Yeah. Yeah, in the rain, that's great! California, that's better. Some of the northern states are very particular, very tight—very strict, because of the snow they get, as in Illinois, Wisconsin, New Hampshire. There's no program around when I got to Nashville. Well, I found out from people from New Hampshire, Wisconsin and Illinois—some of them go through 40 hours of professional / [formal] training. They have safe driving. They better, because that's what they are going to be up against six months out of the year.

That's great Patrick, good news. This is about Rwanda, because I wanted to included other countries to show how we fare here, not just in Nashville, but I want to have a local and a regional and a national look at how much training people get: where they get it, when they get it, and where are the deficient areas. Pinpoint the deficient areas. Then just give that data to the public in the book and let others cogitate on that and figure out what they have to [do]: go into a big conference or they have to write their senators and so forth. I'm not going to do that for them. I'm solely presenting the data, which is in no small measure available via the Internet, how many deaths, how many collisions there are on a daily, weekly, yearly basis. But that doesn't tell the story to why that happens. What tells the story is what you are telling in this interview. (And many others, from Nashville, Wisconsin and from Connecticut New Hampshire, Illinois and from California tell me.)

(On July 17th, 2017 I passed two drivers in the fast lane of the Interstate within a minute or two of each other that were talking on a cell phone and going slower than the traffic in any of the other lanes. I repeat, talking on a cell phone, which is of course somewhat a distraction, and it seemed making them go slower—like they're being more careful, when they should be going faster, and not talking or at least faster than the other lanes to the right. That as we see posted so often IS THE LAW. In Tennessee slower moving traffic must move over to the right.)

Here then are some sobering facts about roadway fatalities and injuries, for those who can and will read them with a view to applying the information for real to their driving practices:

(From one of my Driver Education Manuals of 2006, *Alabama Safety Institute*, Driver Education Course)

Note: similar statistics, in the main exist for the majority of states, and though dated 2006, the statistics, if at all, are relatively higher for recent years.

Just the sheer number of deaths and injuries should tell us something is amiss definitely and needs re-evaluation towards training and education of drivers, universally. When we see that there are *accident prone*[6] drivers on the road driving, and there are a percentage of drivers who elect *unwisely* a fatal collision as a method of solving a life or relationship gone wrong, we are still left with the overwhelming evidence that all drivers need to be better trained, and that even well-trained drivers need to continue their education and training as well. Nothing remains in a constant state for long.

For Alabama, 2006 (In summary)

---

[6]-Accident Prone: one who causes accidents in his/her vicinity, sometimes by their mere presence.

Persons Killed 1,208 up from previous year, 2005 by 5.2 %

Person Injured 43,028 down from previous year, 2005 by 2.6%

Reported Crashes 139,731 down from previous year, 2005 by 2.9%

Miles Traveled 60,394,000,000 up from previous year, 2005 by 1.2%

For myself the Miles Traveled and Reported Crashes are very much secondary statistics. But the other two are horrendous for even an entire state: when we consider that the bulk of humanity in that state, the total driving population *minus the drivers injured and killed* far outweigh in numbers of killed and injured. How, we must ask ourselves can so many avoid collisions, but a much smaller percentage don't? This is the effective question that needs to be asked and answered truthfully.

There is a motorcycle training school in Southern California that well trains motorcyclists to confront and successfully handle every possible scenario that can occur on roadways. The success of this school and its graduates is unimpeachable. It has been in existence more than thirty years. If motorcyclists can be educated and trained with this much success, knowing that they are three to five times more vulnerable than automobile drivers shouldn't we be able to formulate the *correct* teaching methods and training to achieve even better results for drivers of cars and SUVs?

Deaths occur in *fatal collisions*. But if there were no, or greatly reduced collisions then there would be greatly reduced *fatalities and injuries. Is this not evident?*

So, the emphasis I see and I champion is to get drivers to get and hone the skills of driving so as to never collide with a person, animal or solid object.

One night as the rain began to fall in Los Angeles I rode out on my motorcycle. I went through a yellow light but then could not avoid a pick-up truck that simply came out of a gas station without stopping 50 feet on the other side into my right-hand lane. Had I gone to and been trained at the *California Superbike School* I would have laid down the bike and skidded around the truck saving myself a compound fractured right leg and a fractured orbital (eye) socket, two days in intensive care, $22,000.00 in medical costs and six months in a cast that was *very uncomfortable*—not to mention losing those months in *no earned* income. In the many years since I have not collided with any person, animal or solid object, although one vehicle has collided with my vehicle, (sideswiped) non-seriously but yet costly in terms of repair costs.

It is well worth noting as did a master educator once did: that there are people who can drive on the road and there are people driving on roadways who can't drive.* What we can conclude is that steps must be taken to get those persons to join the ranks of the others who can drive. Just to be sure when we say can't drive, we mean the person cannot control a vehicle, which takes them out of the class of *driver*, and puts them in the class of *murderer*. (Potential and real.)

* Not just DUI drivers.

The DUI Driver

(How To Remedy)

The way it's going to happen: the particulars have to be part of the case in front of the judge. If it is not, when the DUI driver gets to the training academy the instructor goes over this whole scenario, finds out what the driver was doing before they drove off and got into a crash or were arrested for DUI. Find out the particulars and replicate the scenario. For example: if the DUI driver was at a party and they were stumbling out of the front door looking for their keys, the instructor does the same thing. Instead of getting in the car, he calls to a friend, "Would someone please drive me home," or "Would you call an Uber or Lyft for me." Then have the offending DUI driver act out the same scenario with sincerity, motion for motion, word for word or almost verbatim. Instructed without critical remarks or demeaning looks by the instructor. The scenario could be altered if there is more than one DUI conviction and drilled until the DUI student can do it with a straight face calmly. Its role playing to accustom a driver who has made a serious blunder to automatically choose the pro-survival method of getting home. This <u>training</u> is done with a sober supportive attitude if it is to succeed. The DUI driver must act with sincerity or he is not passed by the instructor. Any emotional flash-back or joker-degrader comments by the DUI driver is met with a "Do It As Demonstrated." There is no reasoning with the recalcitrant student, because they are below understanding until they can replicate the correct method. THEN, they are in a position to understand why it is the correct way, and why it should be done that way. It may take in some cases several guided run-throughs of one scenario and/or a combination of run-throughs if the DUI student has been convicted of multiple DUI violations.

The attitude towards the student DUI driver is paramount. The DUI driver must perceive that they are in a "learning" position not a disciplinary one. That has already been established by court room appearances, fines and attorney fees. Let the DUI student "blow off steam," if it occurs. It is normal to feel compunction for endangering or injuring the lives of others; but it won't come *usually* without some explosive emotion beforehand. Let the emotional confusion blow off and it will subside. The DUI student will have a realization as well. And that is the guarantee that DUI driving for that student has come to a final end.

We would be wise to also follow this pattern of re-education, refamiliarization of the responsibilities of drivers who are cited and found guilty of texting while driving. The very fact of texting while driving is not itself an immediate censure, if it is done in the following manner: I place my cell phone in a dock attached firmly to the left most area of the windshield just below eye level. And, <u>then</u> I use the facility of dictating a message while I am looking straight ahead at the traffic thus:

Never do we permit fingers to be texting, or to allow drivers to *verbally* text in any other position that obstructs their vision from the traffic in and around them as they are traveling in a vehicle.

Of course the ideal scenario is this, or the driver has a blue tooth set-up whereby they can dictate, hands free, commands and messages via that device while keeping their eyes focused ahead.

SUMMATION & CONCLUSION

Summation:

Here's an observation many may have overlooked. On an eight-lane highway (four lanes going each direction) there are only six *useable* lanes, in reality. Why? Notice that the tractor-trailer drivers stay in the next to right most lane. Why? To avoid having to move back over to that same lane each time another vehicle slower moving at first attempts to merge with the lane they would be traveling in, if they remained in that outer right lane. This is much more the case in rush hour traffic within city limits as the exits are close together and the occurrence of merging and exiting traffic more pronounced.

Why is the innermost lane considered the *fast lane, and is designated the passing lane*? Elementary, almost all exits on fast moving highways (Interstates) begin on the right lane next to the shoulder. This then, is the lane for slowing down to exit. Consequently, the lane left of that lane is moving faster by default. Expand this into a three or four lane (going one-direction) highway and it is evident that the lanes cooperate with each other to allow slower moving vehicles to move to the lane to the immediate right. And, in many States, this is made a state law and is posted frequently on Interstates along the roadway. The fastest lane is the one where there would be no exits, except in rare occasions, the farthest left lane. This lane is designed and there for *passing slower moving vehicles*. Not for traveling in for a great distance as we often see. Otherwise, we get the scenario where a driver passes on the right, the slower moving traffic, into and causes that lane to divert or slow down as he or she barges in at a faster speed. *This practice is also one of the main causes of rush hour slow-down and delays, i.e. a driver without much notice or insufficient notice decides to "get around" another slower or slowing driver and starts a chain reaction winding up all of the lanes in his direction suddenly braking.*

Conclusion, mine: other countries do and require more professional training than the U.S. Consequently, they have lower per capita death rates and lower injury rates—even with more strenuous road circumstances, ex: Rwanda, Philippines. We pay a tremendous amount of dollars annually for auto insurance in the United States. **What would happen if a small fraction of the billions spent on auto insurance and hospitalization was spent per capita (per driver) on professional driver training?**

Our states could mandate that *instead of sending traffic violators and accident causing drivers to traffic school, <u>they are ordered to driver training classes, and</u>* <u>**shown how to do correctly what they failed to do— and passed only when they do it as well**</u>*?* I don't believe joy of causing pain is the reason we go out and get stinking drunk and then drive our vehicle into another one, mangling the bodies of those therein, nor the reason we speed down poorly lit highways and roads at speed limit plus 40 mph to the result of unrecognizable wrecking yard debris. No, I believe we do these things simply because our ability to create life has gone astray so badly that the only solution is to <u>try to</u> uncreate it with force. Notice, *try to* underlined. It never quite uncreates it.

But this work is not aimed at that segment of motorists described above anyway. They are another *different problem,* different than the one this work intends to quell.

Today September 15, 2017, I was reminded by a woman from the United Kingdom of another annoying, unnecessary driver fault: turning out or left to turn right at a corner or driveway. This she reveals is maddening, and I'll add to that that it causes unnecessary stoppage and or stress on drivers behind who do not predict what is going to happen. This phenomenon is widespread and not indigenous to Tennessee drivers. The main type of vehicle and driver seen doing this *habitually* is the SUV. My concept and *recommendation:* Take the turn slower or if it is a driveway go over the curb, it isn't going to damage your SUV, unless there is a culvert that a narrow driveway *bridge* is used to span it. That's an uncommon situation with most driveways.

And, one of my irritations on surface streets: a driver comes to a stop for a jaywalking pedestrian, coming from the other direction traffic lanes: already putting himself in danger. No driver cognizance of vehicles behind, just stops, and in sight of a well painted crosswalk the pedestrian could see and should have used.

It is or should be of interest to all drivers to take note that a typical SUV in the United States weighs *in excess of 1 and a half to 2 times the weight of a passenger car.* For example, the Nissan *Armada,* comes in at a whopping 7,500 lbs. A Nissan *Altima,* weighs at maximum 3,342 lbs. Get where were going with this? If an SUV driver continues to <u>only use the brakes but not gain the skill to</u>

also use the gears to slow the vehicle down, not only will there be additional unnecessary brake repairs, there will be an unending scenario on highways and roadways of seeing these vehicle's (and others) stop lights go on unnecessarily over and over again.

Also, choosing the right gear to travel in improves fuel efficiency—something all drivers have a vested interest in.

Final Conclusion: If we as a nation were to adopt a standard of driver training much like Mr. Pitner offers, with a minimum of 30 hours of professional driver training, with more rigorous testing before granting license, we would see in a short period of time a dwindling of *accidents*, and the resultant carnage and wreckage we have become inured of over the years. That is the hope and future look of this work.

As a final footnote, it is no coincidence that the states with the highest number of hours of supervised driver training, HAW, MA, MD, NH, NJ, have the lowest per capita vehicle fatality rates of all 50 states.

Their relative per capita ratio of vehicular deaths to population, 2016:

HAW   109/1,400,000 = 0.000078 or 0.008% (8 thousandths percent)

MA	359/6,800,000=	0.005%
MD	477/6,000,000=	0.008%
NH	130/1,340,000=	0.009%
NJ	569/8,900,000=	0.006%

States (as example) with minor supervised training requirements:

TX	3,407/27,800,000=	0.012%
FL	2,933/20,600,00=	0.014%

Statistic: More than 25% of teen fatalities on highways are caused by teens in rear seat distracting driver. Shouldn't we be drilling young drivers to pull off road, stop the car and exit the boisterous teens? Shouldn't this be as a matter of rule in driving school (high school, professional) courses a mandatory requirement? 58% are distracted teens (all distractions counted) Source: *National Organizations For Teen Safety.*

Curious fact: there is a 10-minute test that identifies *accident prone* individuals that are the cause of many highway collisions, outside of but not necessarily exclusive of DUI drivers. Shouldn't we d*emand* that this test be administered to all applicants for driving privileges? We go to great lengths to guard against terrorists boarding U.S. commercial planes and other conveyances, government buildings, court houses etc. why can we not demand and screen those that would cause deadly accidents?

Don't just drive defensively, drive *expertly*. Fare thee well.

<u>Look for</u> *How To Stay Healthy Until We Die,* a primer for youth so they can enjoy a healthier life when they attain adulthood. Coming 2018/2019.

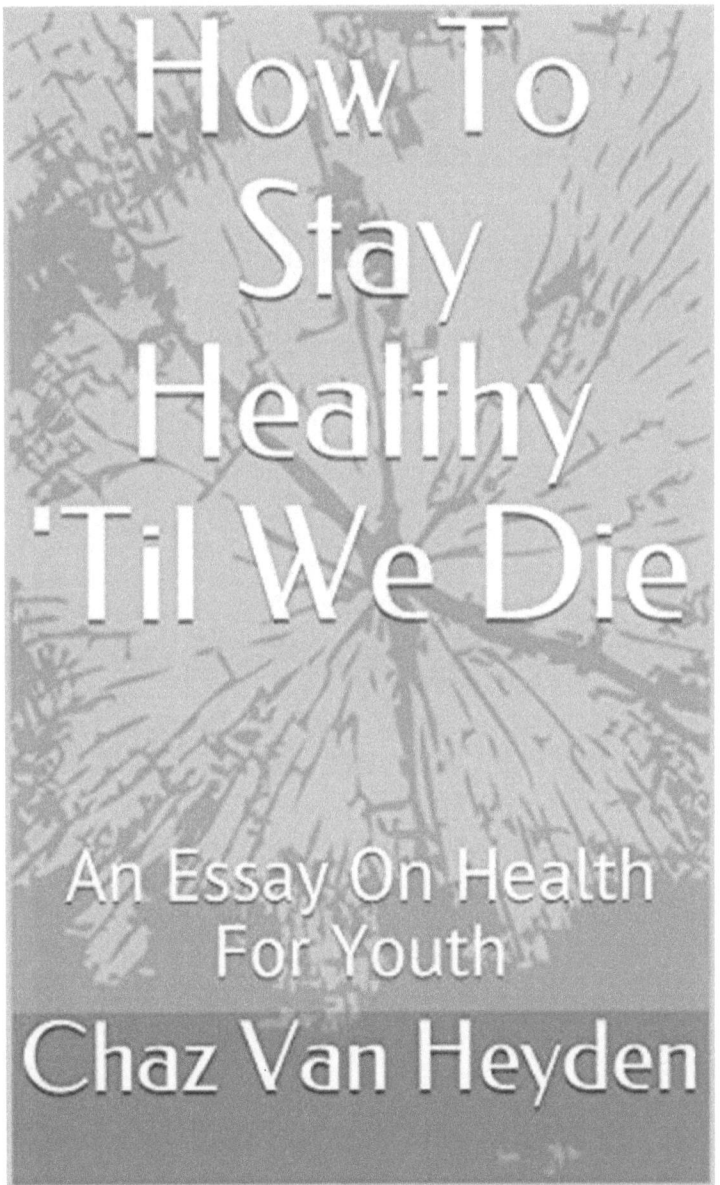

www.ingramcontent.com/pod-product-compliance
Lightning Source LLC
Chambersburg PA
CBHW021419210526
45463CB00001B/441